Homeowner's Guide to Fastening Anything

Robert Scharff

Ideals Publishing Corp.

Milwaukee, Wisconsin

Table of Contents

ISBN 0-8249-6120-X

Published by Ideals Publishing Corporation
11315 Watertown Plank Road
Milwaukee, Wisconsin 53226

Editor, David Schansberg
Cover photo by Jerry Koser

Acknowledgements: I would like to thank my associates at Robert Scharff & Associates, Limited and the various manufacturers for their wonderful assistance. I would also like to thank the National Retail Hardware Association for help provided in suggesting the fasteners most readily available to the home craftsperson.

Putting It Together with Nails

Nails are probably the oldest of all modern-day fasteners, having been used since ancient times for constructing all kinds of wood structures. There is good reason for this long-lasting popularity; nails are the most practical means for fastening pieces of wood together easily, quickly, and inexpensively. Screws are stronger, glue is neater, and either combined with correct joinery becomes stronger still. But nearly all the homes of the nation are fastened together with nails because they are quicker and easier to handle.

Today, there are probably as many kinds of nails on the market as there are projects which require them. You should be able to find the size and shape that is best to use for the job you are doing. Choosing the right nail and driving it properly makes the difference between a quality job and one that falls apart.

Selecting Nails

Nails achieve their fastening or holding power when they displace wood fibers from their original position. The pressure exerted against the nail by these fibers as they try to spring back to their original position provides the holding power. Nails vary according to: size (length and diameter), type of head, type of point, finish of the shank, and material and finish used.

Nail Sizes

The lengths of the most commonly used nails are designated in inches and also "penny" size, a term which originally related to the price per hundred, but now signifies only length. The abbreviation for the word penny is the letter d. Thus, the expression a 2d nail means a two-penny nail. As the wire gauge numbers go down, the penny size number and the diameter of the nail go up.

Nails larger than 20d are frequently called spikes and are usually designated by their length in inches. Some special-purpose spikes that are used in the railroad and boating industry or for fastening wood gutters are larger than 60d but do not carry a penny-weight designation. Nails smaller than 2d are designated in fractions of an inch.

Nail diameter generally increases with length; a 6-inch common nail is nearly four times the diameter of a 1-inch nail. Special-purpose nails, however, may come in only one size or in several lengths but only a single diameter, depending on their purpose.

The minimum diameter of a nail is that which is required to drive the nail satisfactorily without buckling the shank. Hardened steel nails, therefore, can be made with smaller diameters than non-hardened steel nails, while aluminum-alloy nails or specially-designed nails with grooved shanks often need to be made with larger diameters than the plain shank nails. The resistance of nails to withdrawal increases almost directly with their diameter; if the diameter of the nail is doubled, the holding strength is doubled, providing the nail does not split the wood when it is driven. The lateral resistance of nails increases as the diameter increases.

There is a simple rule to follow when selecting nail lengths for both rough (framing) and finish (trim, cabinets) carpentry. The rule applies to hardwoods and softwoods. In the case of hardwoods, the nail penetration into the bottom piece should be one-half the length of the nail. For softwoods, the penetration into the bottom piece should be two-thirds the length of the nail. When edge nailing, regardless of the wood type, be sure that two-thirds of the nail is in the thicker piece. The thickness of the top piece determines the required nail length.

When plywood is used, nail size depends on the thickness. For ¾-inch, use 8d. For ⅝-inch, use 6d or 8d. For ½-inch, use 4d or 6d. For ⅜-inch, use 3d or 4d. For ¼-inch, use ¾- or 1-inch brads, or 3d nails. A 6-inch spacing is about right for most work.

Nailheads

The selection of the nailhead is governed by its end use. The nail may be practically headless, as in the case of a finishing nail, or it may be countersunk to accommodate the nail set when the nailhead is to be set slightly below the surface of the work.

When exceedingly strong fastening is required, a large head is needed. The larger the pull or the softer the material being nailed, the larger the head must be. Thus, when applying asphalt roofing, use broad flathead roofing nails to keep water from seeping

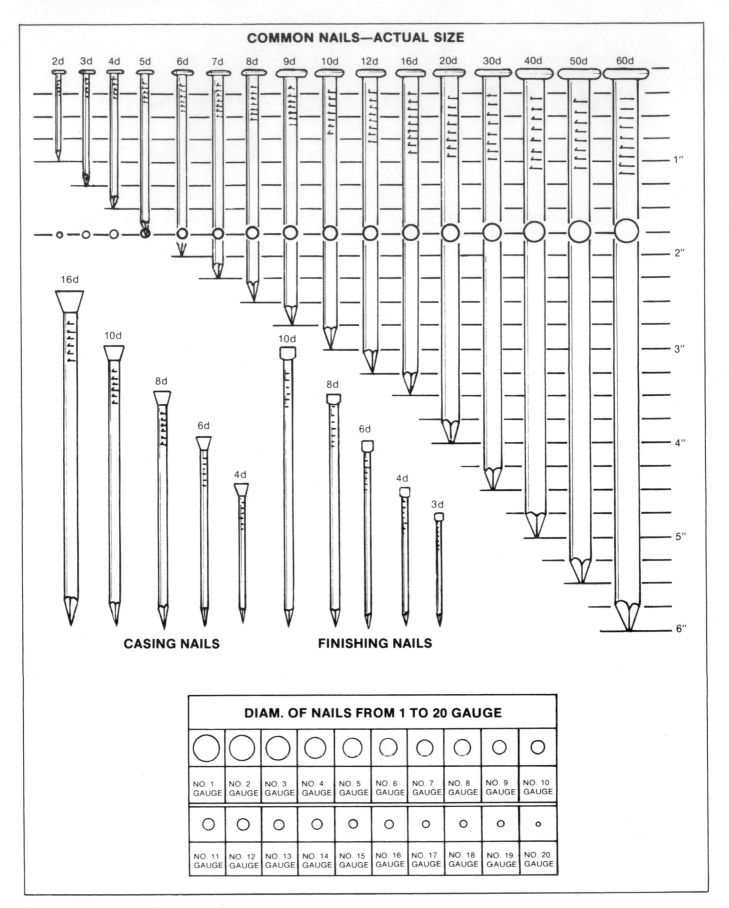

Common nails—actual size.

beneath the head and through the roof. Some roofing nails have a neoprene washer that spreads out when the nail is driven to seal the nail opening, thus preventing water from penetrating.

There are many other special types of heads: round, oval, button, cup head, and so on. Some nailheads are numbered or lettered, which allows for special identification. There are others that have sinker heads, which are ideal for the installation of siding. Projection head nails are another special type that generally are used when acoustical tiles with perforations have to be face-nailed; the lowered head holds fast against the tile at the bottom of a hole while the upper or driving end of the nail is flush with the face of the tile. This eliminates the need for countersinking each nail.

Some nails even have two heads. The lower head, or shoulder, is provided so that the nail may be driven securely home to give maximum holding power while the upper head projects above the surface of the wood to make its withdrawal simple. The reason for this design is that the double-headed or duplex nail is not meant to be permanent. It is used in the construction of temporary structures such as scaffolding and staging and is classified for temporary construction.

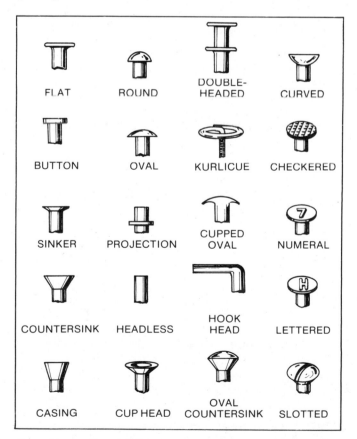

Some examples of different types of heads used for nails. Nails most commonly used by the handyman are flat and casing.

Nail Points

The selection of the nail point is also dependent upon the use of the nail. The point influences both the nail-holding power and the splitting resistance of the wood into which the nail is driven. Most nails have regular diamond points which will allow the nail to hold well without splitting less dense woods during driving. When the nails have long diamond points, such as drywall nails, they are used in particular areas for easy penetration without causing the material into which they are being driven to crumble.

Dense woods should be nailed with blunt-pointed nails to prevent splitting. If not readily available, you can blunt the point of any nail by setting the head on an anvil or any metal block and tapping the point gently with a hammer. Or, the nail can be filed lightly. Blunt-pointed nails actually have increased holding power and are less likely to split the wood, particularly when driven in near the edge.

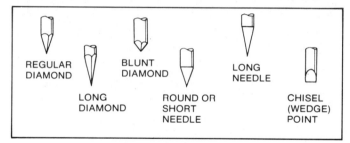

Popular styles of nail points.

Round needle-point nails are used when nailing composition wallboard products to wood since they do not split the material but spread it apart. Long round-point nails are generally used to tack carpeting or fabric in place.

Chisel-point nails are used with hardwoods for they can be driven with ease, especially when using large diameter nails. This point type also allows deep penetration for firm anchoring of heavy objects.

Nail Shanks

The shape of the shank plays a very important part in the nail's holding power. Most common nails are made with smooth round shanks. But this type of shank gives the least holding power. Triangular- and square-shank common nails have increased holding power and are used to fasten furring and flooring to hard surfaces. For example, the triangular cut type is frequently employed to blind-nail hardwood flooring through the edges without splitting. Square-shank concrete nails are used to fasten furring strips and brackets to concrete surfaces. Cut and square-shank nails, however, have become less important to the home handyman in recent years because of the development of special nails for the same purposes.

Barbed nails, which have horizontal or herringbone indentations in the shank, hold better than a smooth nail of the same size, but far less than a threaded nail of equal size. Barbing the nail shank will decrease its resistance to withdrawal if the nail is pulled immediately after driving. However, barbing is advantageous when moderate moisture changes occur in the dry wood into which the barbed nail was driven.

Threaded shanks provide the best holding power and performance. They are made in three basic forms. The first and most common style is the one made with spiral (helical) threads. Often called screw or drive nails, they turn when driven—much like wood screws. They are designed to drive into hardwoods and dense materials. This thread makes an aluminum nail easier to drive.

The second type of threaded nail is the annular-threaded or ringed nail which has a series of closely spaced grooves around its shank. Since this style of thread does not spiral, the nail does not turn as it is being driven. Instead, the wood fibers are forced over the ring shoulders into their annular grooves like wedges. Because of the piercing action of these

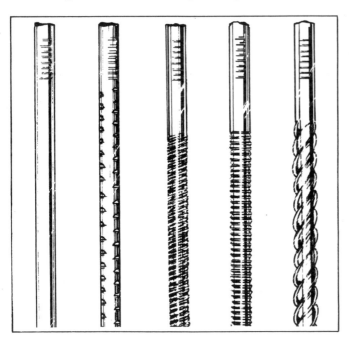

Different types of nail shanks (left to right): standard plain or smooth-shank; barbed-shank; spiral-threaded shank; annular-threaded shank; and knurled-threaded shank.

nails, they can be driven into knots. Annular-threaded nails also give the greatest holding power in soft or medium woods. In addition, they help prevent squeaks in floors and stairs, and help tighten roofing and siding joints or laps against weather. The name of each annular nail usually indicates its use.

Some manufacturers are now producing a threaded nail that combines the wedging features of the annular-grooved nails with the uniform wedging and turning action of the normal spiral-threaded nails. This offers the advantages of both styles.

The third type is the knurled-threaded nail. It has a vertical thread for driving into cinder block, mortar joints, or other relatively soft masonry. It cuts the masonry to minimize cracking and provides good holding power.

Threaded nails are difficult to remove once they are driven in. Although the cost of threaded nails is more than twice that of the same size smooth-shank type, they have much greater holding power (sometimes up to ten times), thus requiring less nails or smaller sizes for most jobs. Cost differentials frequently are outweighed by the many advantages of threaded nails.

Nail Materials and Finishes

Though nails made of steel are the most commonly used, they also can be made of aluminum, stainless steel, copper, brass, and bronze. Nails made of metals other than ordinary steel are designed for fastening objects of the same metal to reduce the corrosive action that occurs when different metals are placed in contact with one another. Aluminum nails, for instance, should be used to secure aluminum gutters to aluminum siding. Of course, they are also suitable for most exterior jobs and can be used with a wide variety of materials, such as wood or asbestos siding and shingles, plastic panels, roofing, gutters and downspouts, porches and decks, and outdoor furniture.

Copper, brass, bronze, and stainless steel nails are also rustproof. In addition, stainless steel nails are resistant to corrosion even when exposed to strong chemicals. Steel nails will rust so they should not be used where rusting would cause discoloration or staining. However, steel nails are available in the following finishes:

The electro-zinc plated steel nail has a shiny but thin coating. It is susceptible to rust, so it is best for interior jobs.

Galvanized nails generally refer to tumbler galvanizing which produces a gray, thin coating. They have better rust-resistance than electro-zinc plated ones, but less than those with hot-dipped galvanized coatings.

Hot-dipped galvanized nails have a high-quality zinc coating with good rust protection.

A temporary finish, cement coating, is a resin coating which makes the nail hold better for a short time. These nails are ideal for box and crate construction.

Blued, heat-treated nails have good temporary rust-resistance but should not be used outdoors.

Bright-finished nails have a bright, uncoated steel finish for use where corrosion resistance is not required.

Quench-hardened nails are heated, quenched, and tempered to increase their resistance to bending when driven into hardwood or masonry.

Colored decorator nails need special treatment; a plastic cap should cover the head of the hammer when driving them.

Common Names of Nails

The commonly used names of nails generally describe the nail's use. For instance, boat nails are special nails used in putting boats together, and railroad spikes are used for holding rails in place. Here is a short description of some of the more popular nail names:

Common Wire Nails and Box Nails Both these nails have flat heads and diamond points, but the diameter of the box nail is somewhat smaller for a given length and penny size. For example, a 10-penny (10d) common nail, which is 3 inches long, has a 9-gauge diameter (0.1483 inch), while a 10-penny box nail, which is also 3 inches long, has a 10½-gauge diameter (0.1277 inch). Thus, if you are

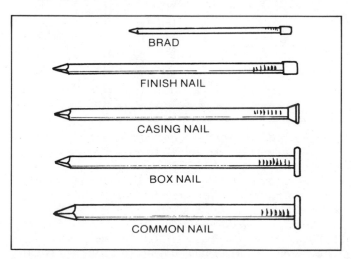

BRAD

FINISH NAIL

CASING NAIL

BOX NAIL

COMMON NAIL

The so-called general purpose nails.

using a common nail that is apparently causing the wood to split, switch to a box nail of an identical penny size. It will probably eliminate the trouble. Both of these nails are used for general structural construction.

Finishing Nails These nails are made from finer wire and have a smaller head than the common nail. They have a cupped head to receive a nail set, so they may be set below the surface of the wood into which they are driven and will leave only a small hole, easily puttied. They are generally used for interior or exterior finishing work and are used for finish carpentry and cabinetmaking. Small finishing nails which measure anywhere from ½ to 1½ inches in length are usually referred to as brads, rather than nails. They are graded by length and wire gauge number and are useful for fine work where little or no stress is involved. When this same type of thin-gauge, light-duty nail is made with a flat head, it is referred to as a wire nail, rather than as a common or box nail.

Casing Nails Though they look almost the same as finishing nails and are used for the same purpose, casing nails, because of their flat countersink heads, may be driven flush and can remain that way. Also, casing nails have slightly heavier shanks to provide increased holding power.

Common, box, finishing, and casing nails are considered general-purpose nails since they are used for general carpentry and construction. Special-purpose nails, as the name implies, are for specific uses. Some of the nails are available in different shank styles. Some of the more popular special-purpose nails are listed below.

Drywall These ring-shanked nails are used for attaching sheets of drywall gypsum board to interior wood wall studs. The flat, slightly countersunk heads permit driving just below the surface to form a depression for spackling. Threaded drywall nails have better resistance to conditions that cause regular drywall nails to pop. Smooth-shanked wallboard nails coated for extra holding power are also available.

Flooring Quench-hardened, spiral-shanked nails, either casing head or countersunk, are ideal for laying tongue-and-groove hardwood flooring. Flooring nails are available in 6d, 7d, and 8d—all 11½ gauge.

Plasterboard These blued, smooth nails with flat heads and long diamond points are used to fasten plasterboard to interior wood wall studs.

Underlayment These bright-finished, ring-shanked nails with either flat or countersunk heads are used for laying plywood or composition subflooring over existing wood floors or floor joists.

Roofing These are round-shafted (smooth, ringed, or spiral), diamond-pointed, galvanized nails of relatively short length and comparatively large heads. They are designed for fastening flexible roofing materials and for resisting continuous exposure to weather. Roof nails are available in lengths from 1 to 2 inches, with head diameters varying accordingly. If shingles or roll roofing is being applied over old roofing, the roofing nails selected must be of sufficient length to go through the old material and secure the new. Asphalt roofing material is fastened with corrosion-resistant nails, never with plain nails. Begin nailing in the center of the shingle, just above the cutouts or slots, to avoid buckling. When nailing asphalt roofing to plywood, the nail should be long enough to go through the plywood. Since roofing nails have a rather heavy shank in proportion to their length, do not nail them into thin lumber because this would cause splitting.

Siding These galvanized nails are used for applying residential wood lap siding to plywood or fiberboard sheathing. Also, consider using threaded nails when nailing shingle or shake siding to a house. They will hold the siding firmly against storms and heavy winds. Barbed asbestos-shingle nails also hold well.

Pallet These threaded nails are good for nailing sheathing, framing, furring strips, trim, fencing, and other general construction or repair work.

Masonry and Concrete These quench-hardened nails, with flat, countersunk heads and diamond points are used in fastening wood to concrete mortar or masonry. Most have knurled threads and are usually available in lengths from ¾ to 3½ inches. Helyx concrete nails are also available in lengths of ¾ to 3½ inches, and because of their heavy, screw-type shank they hold well in concrete, mortar, and cinder blocks. Because of the high resistance of

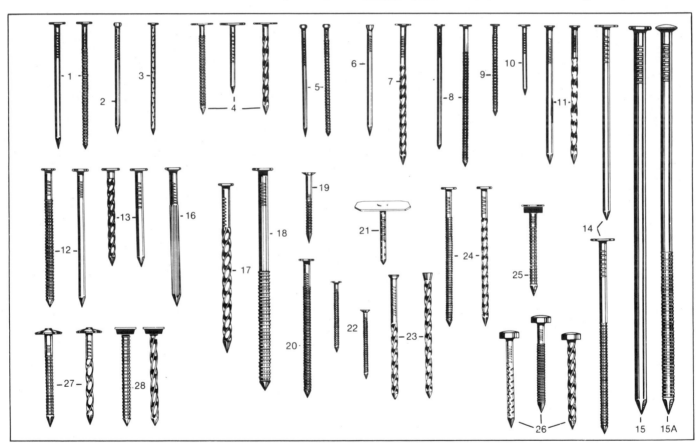

The more popular special-purpose nails. For exterior applications (hot-dipped zinc-coated): (1) Wood Siding, Box (Plain & Annular); (2) Flooring; (3) Insulating, Plastic Siding; (4) Asphalt Shingle (Annular, Plain & Knurled); (5) Cedar Shake (Plain & Annular); (6) Casing; (7) Cribber; (8) "Split-Less" Wood Siding (Plain & Annular); (9) Asbestos; (10) Cedar Shingle; (11) Hardboard Siding (Plain & Knurled); (12) Common (Annular & Plain); (13) Aluminum, Steel & Vinyl Siding (Knurled & Plain); (14) Insulation Roof Deck (Plain & Annular); (15) Gutter Spike (Plain); (15A) Gutter Spike (Annular). For interior and other nails: (16) Masonry; (17) Pole Barn, Truss Rafter (Knurled); (18) Pole Barn (Annular); (19) Drywall; (20) Underlayment, Plywood (Subfloor; sheathing, etc.); (21) "Square-Cap" Roofing; (22) Underlayment (Flathead & Countersunk); (23) Flooring (Nail & Knurled); (24) Pallet (Annular & Knurled). For metal roofing nails: (25) Rubber Washer; (26) Compressed Lead Head (Barbed, Annular & Knurled); (27) Umbrella Head (Annular & Knurled); (28) Lead Washer (Annular & Knurled).

masonry to penetration, concrete and masonry nails must be made of a high grade, heat-treated steel increasing the possibility of breaking, chipping, or shattering, and the possibility of injury. Safety glasses should always be worn and driving directions carefully followed.

1. Start the nail perfectly straight. Hit it squarely with a tapping, one-two stroke.
2. A heavy ball peen hammer with a large striking face or a hand drilling hammer should be used when driving hardened nails. Nail hammers, designed for driving relatively soft common and finishing nails, should never be used to strike hardened nails.
3. Do not use a rebound stroke; allow the hammer to lay on the head of the nail at the finish of the stroke. A heavy stroke is not required.
4. Do not drive more than ¾ inch into the masonry.
5. Any nail longer than ¾ inch must be driven through wood or other supporting material before penetrating the concrete.

Escutcheon Pins These small brass (or stainless steel) nails with round heads are available in lengths from 3/16 to 2 inches, in gauges sizes 10 to 24, with smooth or annular-threaded shanks.

Panel These threaded-shank brads are usually available in lengths of 1 and 1⅝ inches and in about 15 colors to match most plywood and hardboard paneling. They have superior holding properties when compared with regular brads and finishing nails.

General purpose nails are usually purchased by the pound. The current trend in merchandising special purpose nails for the home handyman is in packages rather than in bulk.

Before Driving Nails

When driving nails, it is important to consider the condition of the wood. If it is soft, the nail will drive in easily, but it will also pull out easily. The harder the wood, the more difficult to drive a nail, but the better it will hold. Hardwoods split more easily than soft. Starting with softwoods like balsa and pine, through rock maple and oak, and on to ironwood and teak, you come to a point where a nail thick enough to be driven without bending is also thick enough to split the wood. Predrilling dense woods can help prevent splitting especially when large-diameter nails are used. The drilled hole should be about 75 percent of the nail diameter. Woods without a uniform texture, like southern yellow pine and Douglas fir, split more than uniform-textured woods

General Purpose Nails

Size	Length	Common Gauge	Common # Per Pound	Finishing Gauge	Finishing # Per Pound	Casing Gauge	Casing # Per Pound
2d	1″	15	845				
3d	1¼″	14	540	15½	880		
4d	1½″	12½	290	15	630	14	490
5d	1¾″	12½	250				
6d	2″	11½	165	13	290	12½	245
7d	2¼″	11½	150				
8d	2½″	10¼	100	12½	195	11½	145
9d	2¾″	10¼	90				
10d	3″	9	65	11½	125	10½	95
12d	3¼″	9	60				
16d	3½″	8	45			10	72
20d	4″	6	30				
30d	4½″	5	20				
40d	5″	4	17				
50d	5½″	3	13				
60d	6″	2	10	# PER POUND IS APPROXIMATE			

such as northern and Idaho white pine, sugar pine, or ponderosa pine. The most common way to reduce splitting is the use of small-diameter nails. The number of small nails must be increased to maintain the same gross holding strength as with larger nails.

The moisture content of the wood at the time of nailing is extremely important for good nail holding. If smooth-shank nails are driven into wet wood, they will lose about three-fourths of their full holding ability when the wood becomes dry. This loss of holding power is so great that siding, barn boards, or fence pickets are likely to become loose when smooth-shank nails are driven into green wood that subsequently dries. Always try to use well-seasoned wood.

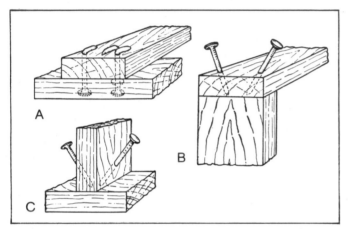

Methods of increasing the holding power of nails: (A) clinching, (B) skewing, and (C) toenailing.

In order to increase their holding power, common nails that are used in rough carpentry work should be clinched. If a portion of a nail extends beyond the surface of the wood, that portion may be bent over. If common nails are clinched, they will have about 45 percent greater holding power than a corresponding nonclinched nail. The clinched portion of a nail should be at least ½ inch long. Longer clinched portions do not provide much more holding power. Clinching should be done perpendicular to the grain of wood.

Whenever possible, nails should be driven at an angle slightly toward each other (skewed). This is especially important when driving into end grain; the slight angle will increase holding power. Nails should be driven through the thinner piece of wood into the thicker one.

In situations where it is impractical to face-nail (as when nailing vertical studs to floors or ceilings), toenailing should be used. This consists of driving nails in at an angle from either side of the stud or beam. It gives increased holding power, particu-

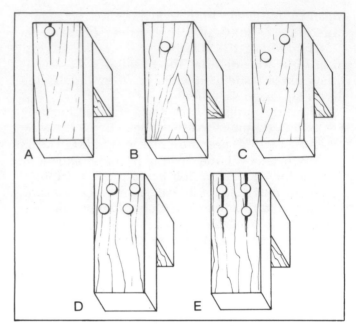

Positioning of the nails is very important. A nail placed too close to the end grain will split out. (A) Placed too near the edges of the board, it will also produce splits. (B) A single nail placed back from the edge will hold without splitting, but it will permit a certain amount of swing in the joint. (C) Two nails, not placed in the same grain line, will eliminate the swing and more than double the strength. (D) When driving more than one nail, it is always wise to stagger the nails so none are in the same line; (E) Otherwise the wood will split. A few nails of the proper type and size, properly placed and driven, will hold better than a great many driven close together.

larly when nails are driven in from opposite sides so that they act as cleats. But, when toenailing, use a nail long enough to pierce through one member being joined and into the other so that about half the length of the nail enters the second member. A 10d nail is generally used when toenailing a 2 by 4, a 16d with a 4 by 4.

Nails should be carefully placed in the work to provide the greatest holding power. When driven with the grain, they do not hold as well as when driven across the grain.

Up to this point in the chapter, the holding power has been considered as the nail's ability to hold against an outward pull. Nails have other holding characteristics, but it is important not to use them for strength alone. Rather than putting the load along the length of the nail, it is best to drive it so that the greatest strain is placed crosswise against the nail. As is quite obvious, nails can easily be pulled out, but it would be very difficult to shear them off. When building anything, remember to keep this in mind.

Nails can be placed in structural joints so that they are driven deeper or have their loads forced against their shear angles. If you come across an area where it is impossible to do this, you will need an additional means of securing the joint. In areas where the nail would be driven into the end grain use the alternate method to prevent splitting the wood and to allow the load to drive the nail further into the stud. Using these suggested techniques, you can be assured of a great deal of safety in your building endeavors.

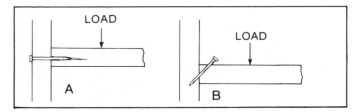

(A) Nail driven into end grain splits wood. (B) Here the load drives the nail deeper into stud.

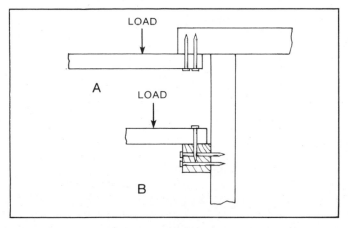

(A) If the load is placed along the length of the nails, they will pull out. (B) Here the top nail is driven deeper and the load is shifted to the two nails placed in the shear position.

Driving Nails

To drive most nails, use a claw or carpenter's hammer. Grasp the handle with the end flush with the lower edge of the palm. Keep the wrist limber and relaxed. Grasp the nail with the thumb and forefinger of the other hand, and place the point at the exact spot where it is to be driven. Unless the nail is to be purposely driven at an angle, it should be held perpendicular to the surface of the work. Strike the nailhead squarely, keeping the hand level with the head of the nail. To drive the nail, first rest the face of the hammer on the head of the nail; then raise the hammer slightly and give the nail a few light taps with a wrist movement to start it. This helps give you proper aim and holds the nail in place during succeeding blows. Never use the cheek

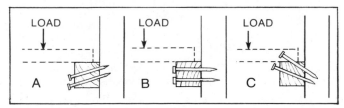

(A) Nails will pull out quickly; (B) Here nails are in the shear position; (C) Placing nails in this manner will cause the load to drive them deeper and will provide better holding power.

of the hammer for driving; the face has been specially processed for striking. Take the fingers away from the nail and continue to drive the nail with firm blows using the center of the hammer face. Nails that do not drive straight or that bend should be pulled out and discarded. If after several attempts the nail continues to bend or go in crooked, make sure you are not trying to drive the nail into a knot or some other obstruction. If you are, drill a small hole through the obstruction and then drive the nail through.

There are times when it may be necessary to start a nail with one hand. This can be done in either of the following ways:

1. Insert the nail between the claws of the hammer, with the head of the nail resting against the head of the hammer. Drive the nail slightly into the wood; then release it from the claw and finish driving in the usual manner.
2. Rest the head of the nail against the side of the hammer and steady it in position with the fingers. Start the nail with a sharp tap of the hammer held in this manner; then finish driving in the usual manner.

A small nail can be held by piercing it through a piece of light cardboard that can be held comfortably.

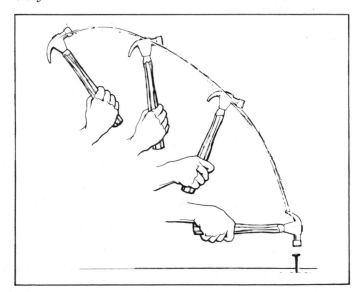

The proper hammer swing.

Recommended Nailing Practices

Proper fastening of frame members and covering materials provides the rigidity and strength to resist severe windstorms and other hazards. Good nailing is also important from the standpoint of normal performance of wood parts. For example, proper fastening of intersecting walls usually reduces wall cracking at the inside corners. The schedule below outlines good nailing practices for the framing and sheathing of a well-constructed wood-frame house.

Joining	Nailing Method	Number	Size	Placement
Header to joist	end nail	3	16d	
Joist to sill or girder	toenail	2	10d or	
		3	8d	
Header and stringer joist to sill	toenail		10d	16″ o.c.
Bridging to joist	toenail each end	2	8d	
Ledger strip to beam, 2″ thick		3	16d	at each joist
Subfloor, boards 1″ x 6″ and smaller		2	8d	to each joist
1″ x 8″		3	8d	to each joist
Subfloor, plywood: At edges			8d	6″ o.c.
At intermediate joists			8d	8″ o.c.
Subfloor (2″ x 6″)	blind nail			
T&G) to joist or girder	(casing and face nail)	2	16d	
Soleplate to stud, horizontal assembly	end nail	2	16d	at each stud
Top plate to stud	end nail	2	16d	
Stud to soleplate	toenail	4	8d	
Soleplate to joist or blocking	face nail		16d	16″ o.c.
Doubled studs	face nail, stagger		10d	16″ o.c.
End stud of intersecting wall to exterior wall stud	face nail		16d	16″ o.c.
Upper top plate to lower top plate	face nail		16d	16″ o.c.
Upper top plate laps and intersections	face nail	2	16d	
Continuous header, two pieces, each edge			12d	12″ o.c.
Ceiling joist to top wall plates	toenail	3	8d	
Ceiling joist laps at partition	face nail	4	16d	
Rafter to top plate	toenail	2	8d	
Rafter to ceiling joist	face nail	5	10d	
Rafter to valley or hip rafter	toenail	3	10d	
Ridgeboard to rafter	end nail	3	10d	
Rafter to rafter through ridgeboard	toenail	4	8d	
	edge nail	1	10d	
Collar beam to rafter: 2″ member	face nail	2	12d	
1″ member	face nail	3	8d	
1″ diagonal let-in brace to each stud and plate (four nails at top)		2	8d	
Built-up corner studs				
Studs to blocking	face nail	2	10d	each side
Intersecting stud to corner studs	face nail		16d	12″ o.c.
Built-up girders and beams, three or more members	face nail		20d	32″ o.c.
Wall sheathing: 1″ x 8″ or less horizontal	face nail	2	8d	at each stud
1″ x 6″ or greater, diagonal	face nail	3	8d	at each stud
Wall sheathing, vertically applied plywood				
⅜″ and less thick	face nail		6d	6″ edge
½″ and over thick	face nail		8d	12″ intermediate
Wall sheathing vertically applied fiberboard				
½″ thick	face nail			1½″ roofing nail
25/32″ thick	face nail			1¾″ roofing nail (3″ edge and 6″ intermediate)
Roof sheathing boards, 4″, 6″, 8″ width	face nail	2	8d	at each rafter
Roof sheathing, plywood: ⅜″ and less thick	face nail		6d	6″ edge and 12″
½″ and over thick	face nail		8d	intermediate

If hardwood resists a brad, a hole can be drilled in the hardwood. Use, as a bit, a beheaded brad or a needle cut above the eye. Dipping a nail in paraffin will help it enter hardwood more easily. Some professional carpenters store paraffin in a hole drilled in the handle of a hammer. Hot paraffin is poured into the hole and allowed to cool.

To pull a nail, slide the claw of the hammer under the nailhead. Pull back on the handle until the handle is nearly vertical; then slip a block of wood under the head of the hammer, and pull the nail completely free. The claw hammer should not be used for pulling nails larger than 8d (8 penny). For larger nails, use a wrecking bar.

Nail Set A nail set is used to set (meaning to countersink slightly below the surface) the heads of nails in finish carpentry. Nail sets are available in various sizes. The purpose of setting is to improve

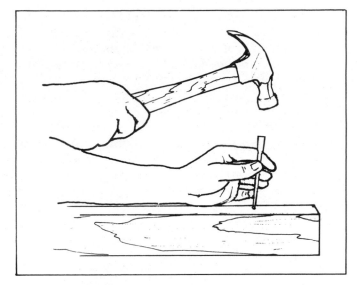

Setting a nail with a nail set.

AMERICAN CUT UPHOLSTERERS	CARPET	BASKET	BILL POSTERS	COPPER CUT	HIDE TACKS	CARPET LAYERS
(incl. Shade and Canvas) 2 to 24 oz.	4 to 10 oz.	4 to 14 oz.	3 to 10 oz.	Flathead Ovalhead 3 to 26 oz.	8 to 26 oz.	8 to 24 oz.

LIGHT TRIMMERS	BARBED WEBBING	GIMP	LACE	ALUMINUM AND COPPER SCREEN	CUPPED HEAD SURVEYORS	BRASS CANOE	LEATHER HEAD CARPET TACKS
1½ to 16 oz.	12 and 14 oz.	2½ to 10 oz.	2 to 14 oz.	3 and 4 oz.	14 oz.	2½, 6 and 12 oz.	8 oz.

Ounce	1½	2	2½	3	4	6	8	10	12	14	16	18	20	22	24
Length	7/32″	¼″	5/16″	3/8″	7/16″	½″	9/16″	5/8″	11/16″	¾″	13/16″	7/8″	15/16″	1″	1⅛″

Standard Counts per Pound

Ounce	1½	2	2½	3	4	6	8	10	12	14	16	18	20	22	24
Carpet	-	-	-	1904	1600	1248	1104	880	-	-	-	-	-	-	-
Upholsterers	7328	5600	4032	3000	2400	1760	1440	1200	1040	800	720	640	576	512	440
Trimmers	8000	6400	5600	4000	3008	2640	1792	1440	1200	1024	896	-	-	-	-
Bill Posters	-	-	-	1264	960	608	544	496	-	-	-	-	-	-	-
Gimp	-	-	5000	3488	2992	2496	1840	1600	-	-	-	-	-	-	-

Styles of cut tacks and sizes available.

the appearance of the work by concealing the nail-heads. A nail is set by placing the tip of the nail set on the head of the nail and striking the set a blow or two with the hammer. To keep from creating too big a hole when using a nail set, hold it firmly in the left hand, between thumb and forefinger, with the hand resting on the wood. Press the nail set onto the head of the nail and strike it, checking its position before each blow. This is particularly important when driving nails in moldings or corners, or when toe-nailing. The small surface hole created above the head of the nail can usually be plugged with putty if necessary.

Tacks

There is a wide variation of cut tacks available to handle anything from carpet to posters to screening. Upholsterers tacks are also available with decorative heads. Standard cut upholsterer's tacks are available in lengths from 7/32 to 1⅛ inches.

Tacks are face-driven in the same manner as regular nails, except a tack hammer is used in place of the claw type. Most tack hammers have a magnetic end for starting the tack, while the opposite end is used for driving it.

Dowel Pins   Dowel pins or sprigs are small

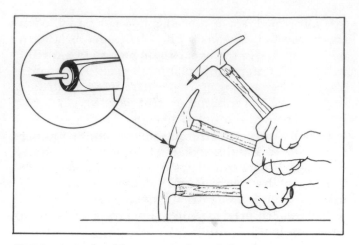

Driving a tack with a magnetic tack hammer.

headless nails with barbed shanks. They are generally available in lengths of ½ to 2 inches, with a No. 8 steel wire gauge or 0.162 inch diameter.

Hinge Pins   These ovalhead nails are occasionally used in place of screws to fasten cabinet door hinges. The light hinge nails are ¾ inch in diameter, while the heavy type are ¼ inch in diameter. Although they are generally available in sizes from 1½ to 3 inches, the heavy type can be up to 4 inches in length. They also come in bronze and brass finishes.

Fastening Wood with Screws and Bolts

When fastening pieces of wood to each other or when anchoring objects to wood surfaces, screws and bolts offer several advantages over nails. If your fastening task falls into one of the following categories, use screws or bolts:

- When better holding power is necessary. This is especially true when the pull of the load is directed along the length of the fastening device; e.g., when hanging a door. The work will not loosen even under repeated vibration and heavy stress.
- When there is a possibility that you may want to take the work apart in the future. It is far easier to remove screws than to pull nails. Furthermore, there is less chance of damaging the work.
- When it is necessary to securely draw together the items being fastened.
- When there is a chance that you will damage the piece of work while applying the fastener. For instance, there is less opportunity for a screwdriver to slip and damage a fine finish than there is for a hammer to slip.
- When a neater or decorative appearance is desired.

If the task you are doing does not fall into one of these categories, it is best to employ nails. They are less expensive than screws and require less time and effort to install.

Wood Screws

Wood screws thread their way into the wood, while wood bolts go through a predrilled hole and are held by a nut threaded onto the ends of the bolts. Wood screws are designated by head types, material, and size.

Wood Screwhead Types All wood screws consist of three different parts: a head, which is slotted so that the screw can be driven into the wood with a screwdriver; an unthreaded body or shank section immediately below the head; and a threaded portion which tapers to a point at the tip. Though screws may vary in many ways, they are most often classified according to the shape of the head. The most popular head types for wood screws are: flathead, roundhead, and ovalhead, with both straight-slotted and recessed screwdriver openings. The most popular screw with a recessed opening is the Phillips screw. This screw has what appears to be two slots at right angles to each other. A conventional slotted screwdriver should never be used to drive a Phillips screw—or any other type of screw with a specialized opening. A recessed head permits you to apply maximum driving pressure when installing the screw, because the special screwdriver makes a close fit with the slots. This allows pressure to be applied on several sides rather than on only two as with a straight-slot head. There is less chance of a screwdriver slipping out of a recessed head and marring the work.

Always match your screws to the job. For instance, use flathead screws where you do not want the head to protrude above the work surface. They can be driven flush with the surface or can be concealed with wooden plugs or wood putty. They also can be used to fasten metal brackets, hinges, mending plates, and other hardware items which have predrilled countersunk holes to receive the screwheads.

The roundhead screw protrudes above the surface in a half circle and, for practical purposes, is not countersunk below the wood surface. This screw is usually used for work that is likely to be disassembled and can also be used for fastening thin materials, such as sheet metal and plastic, to wood.

The ovalhead screw is somewhat of a compromise between the flathead and the roundhead screws. The bottom of the head is tapered so that the screw can be sunk below the surface, but the top of the head is oval to protrude above it. Since ovalhead screws are quite decorative, they are generally used where good appearance and easy removal are desired.

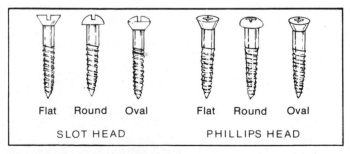

The three common screwheads.

16 Fastening Wood with Screws and Bolts

Wood Screw Materials The common wood screw is usually made of unhardened steel, stainless steel, aluminum, bronze, or brass. The steel may be bright finished or blued, or zinc, cadmium, copper, or chrome plated. The various finishes are usually employed as follows:

1. Aluminum—rustproof—use with aluminum applications.
2. Blued or unfinished steel—inexpensive—will rust.
3. Brass—rustproof—use for exposed appearance.
4. Cadmium or zinc plated—general use.
5. Chrome or nickel plated—use for exposed appearance.
6. Copper plated—use for antique appearance.
7. Silicon bronze—marine use.

Steel screws are generally stronger than brass, bronze, or aluminum, but unless they are specially plated to resist rust, the brass and aluminum screws are better for use outdoors where rust may be a problem. Remember that cost, special purpose application, and material being held will also help to determine the selection of the material to be used. For example, brass screws are always used with oak. If steel screws are used, acid in the oak may make stains. When installing a brass screw, which is relatively soft, be careful not to break it off as it is tightened in place.

Wood Screw Sizes

The length of a flathead wood screw is the overall length, but the length of round- and fillister-head screws is measured from the point to the underside of the head. The length of an ovalhead screw is measured from the point to the edge of the head.

Standards for screws have been established by cooperation between the manufacturers and the United States Bureau of Standards so that standard screws of all screw manufacturers are alike. The size of an ordinary wood screw is indicated by the length and body diameter (unthreaded part) of the screw. They come in lengths which vary from 1/4 inch to 6 inches. Screws up to 1 inch in length increase by eighths, screws from 1 to 3 inches increase by quarters, and screws from 3 to 6 inches increase by half-inches.

Body diameters are designated by gauge or shank numbers, running from 0 (smallest) to 24 (largest). The gauge is an arbitrary number that represents

Screw Sizes and Dimensions

Gauge or Shank Size Numbers

Actual Length (inch)	0	1	2	3	4	5	6	7	8	9	10	11	12	14	16	18	20	24
1/4	x	x	x	x														
3/8	x	x	x	x	x	x	x	x	x	x								
1/2		x	x	x	x	x	x	x	x	x	x	x	x					
5/8		x	x	x	x	x	x	x	x	x	x	x	x	x				
3/4			x	x	x	x	x	x	x	x	x	x	x	x	x			
7/8			x	x	x	x	x	x	x	x	x	x	x	x	x			
1				x	x	x	x	x	x	x	x	x	x	x	x	x	x	
1 1/4					x	x	x	x	x	x	x	x	x	x	x	x	x	x
1 1/2					x	x	x	x	x	x	x	x	x	x	x	x	x	x
1 3/4						x	x	x	x	x	x	x	x	x	x	x	x	x
2						x	x	x	x	x	x	x	x	x	x	x	x	x
2 1/4						x	x	x	x	x	x	x	x	x	x	x	x	x
2 1/2						x	x	x	x	x	x	x	x	x	x	x	x	x
2 3/4							x	x	x	x	x	x	x	x	x	x	x	x
3							x	x	x	x	x	x	x	x	x	x	x	x
3 1/2									x	x	x	x	x	x	x	x	x	x
4									x	x	x	x	x	x	x	x	x	x
4 1/2												x	x	x	x	x	x	x
5												x	x	x	x	x	x	x
6														x	x	x	x	x
Threads per inch	32	28	26	24	22	20	18	16	15	14	13	12	11	10	9	8	8	7
Diameter of screw (inch)	.060	.073	.086	.099	.112	.125	.138	.151	.164	.177	.190	.203	.216	.242	.268	.294	.320	.372

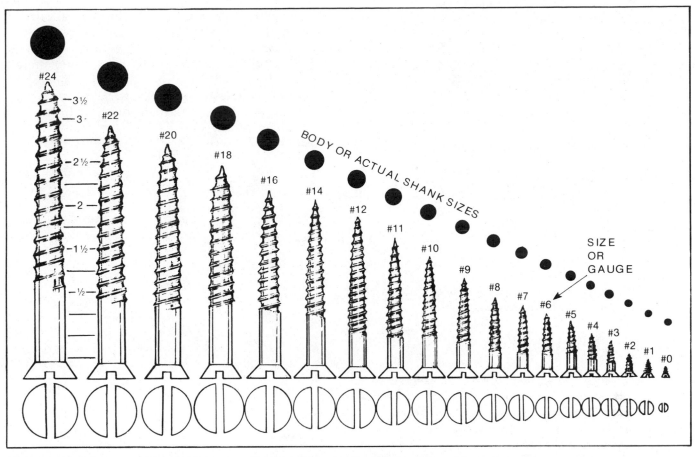

Body and head sizes of standard screws. Sizes are also applicable for recessed opening screws.

no particular measurement but indicates relative differences in the diameter of the screws. In fact, the gauge number can vary for a given length of screw. For instance, a 1-inch screw is available in gauge numbers of 3 through 20. The No. 3 would be a thin screw, while the 20 would have a large diameter. From one gauge number to the next, the size of the wood screw changes by 13 thousandths (.013) of an inch. Using the smallest gauge (No. 0) as a base, you can find the body diameter of any screw number. For example, a No. 5 screw is .125 (.060 + .013 × 5), or ⅛ inch. It is interesting to note that the screw gauge numbers and wire gauge numbers of nails are opposite; that is, a screw gauge number increases as the diameter increases, while the wire gauge used for nail diameters has higher numbers for smaller diameters. Above a No. 12 screw, only even gauge numbers are considered standard.

A screw gauge is a measuring device used in order to determine the numbered size or gauge of the screw. The head of the screw is inserted into the V slot slipped into the groove until the size is noted along the scale. A screw gauge also has a ruled edge for measuring the length.

Wood screws are usually priced and sold by the box. The designation of length, gauge number, and type appears on the container. For instance, if the box is labeled "1½-8-F.H.," it means a No. 8 flathead screw, 1½ inches long. Screws 4 inches and shorter are factory packed by the gross. Those over 4 inches come in packages of one-half gross.

As a general rule, the length of a screw for holding two pieces of wood together should be such that the body extends through the piece being screwed down so the threaded portion will then enter the other piece. The wood screw simply passes through the hole in the top piece and the threads take hold in the bottom piece. Select the screw length so that at least two-thirds of it are in the base material to which you are fastening. The threaded portion can be as long as necessary, but it should always be at least ⅛ inch less than the thickness of the bottom piece. Choose a small diameter screw for thin woods, a larger diameter for heavy woods. Thicker (larger diameter) screws naturally have more strength than thinner ones, so the selection of size will have to be based on the strength needed in the joint, as well as on the thickness of the wood being joined.

When selecting a screw, remember that, as with nails, the maximum area of wood fiber gripping the screw provides the maximum holding power. The power referred to is in an outward direction, oppo-

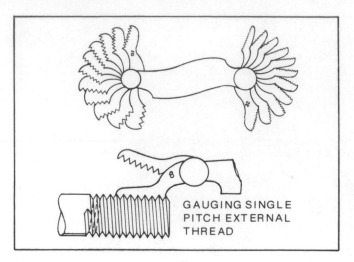

GAUGING SINGLE PITCH EXTERNAL THREAD

A typical screw pitch gauge and how it is used.

site that in which the screw point is aimed. Outward pull against the threads is holding power.

Before Driving Wood Screws

The most common error in driving a wood screw is applying too much pressure on the screwdriver when starting the screw. Without a pilot or starting hole, ordinary screws tend to follow the grain of the wood and can be difficult to drive straight. Providing a pilot hole eliminates this problem and prevents the wood from splitting, especially near the edges and ends of the stock. If the screw is a small one, use a brad awl or screw-hole starter tool. Larger screws require pilot holes made with bits or twist drills.

1. For softwoods (pine, spruce, etc.), drill a hole only half as deep as the threaded part of the screw.

2. For hardwoods (oak, maple, birch, etc.), drill the hole as deep as the screw.

3. For large screws being driven into hardwoods, follow this procedure: First drill a pilot hole which is slightly smaller than the threaded part of the screw. Use a second drill to enlarge the hole at the top; this drill should be the same diameter as the upper (shank) portion of the screw.

To simplify the task of drilling two different diameter holes into a joint when the pieces of wood are already assembled, you can purchase special two-stage drills which automatically bore the right size pilot and shank hole in one operation. They also cut a recess in the surface at the same time for counter-

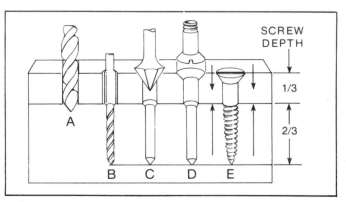

Steps in sinking a flathead screw: (A) drill the clearance or shank hole; (B) drill the pilot hole; (C) countersink; (D) check amount of countersink with the screw; (E) check that screw is properly set.

Safe Load per Inch of Threads					
Screw Number					
	4	8	12	16	20
In oak (hardwood)	80 lbs.	100 lbs.	130 lbs.	150 lbs.	170 lbs.
In yellow pine (softwood)	70 lbs.	90 lbs.	120 lbs.	140 lbs.	150 lbs.
In white pine (softwood)	50 lbs.	70 lbs.	90 lbs.	100 lbs.	120 lbs.

Example: A No. 20 screw set with threads 1 inch into the cross grain of oak will hold 170 pounds; set into white pine, it will hold 120 pounds; set into the end grain of white pine 1 inch, it will hold 72 pounds; but if set into the end grain of white pine only ½ inch, it will hold only 36 pounds. To increase load limits, use additional screws. When the screw is set with the grain, use 60 percent of the indicated load figure. Setting into end grain reduces holding power 40 percent.

Table of Screw Sizes and the Drill Gauge Numbers to Be Used for Maximum Holding Power			
	Pilot Holes		**Shank or Clearance Holes**
Screw No.	**Hardwoods**	**Softwoods**	
0	66	75	52
1	57	71	47
2	54	65	42
3	53	58	37
4	51	55	32
5	47	53	30
6	44	52	27
7	39	51	22
8	35	48	18
9	33	45	14
10	31	43	10
11	29	40	4
12	25	38	2
14	14	32	D
16	10	29	I
18	6	26	N
20	33	19	P
24	D	15	V

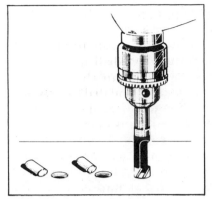
Cutting wooden plugs.

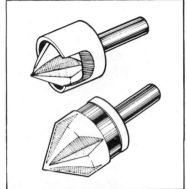
Typical countersinks.

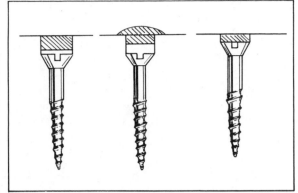
Methods of concealing screwheads.

sinking flathead screws or for counterboring any type of screw.

To completely conceal the screwhead, use a piece of dowel rod or a wooden plug. Plugs can be cut with a tool called a plug cutter, which fits into an electric drill. Plugs should be cut from the same kind of wood as that in which they are to be inserted, and the grain should match as closely as possible. They should be cut so that the grain runs across the plug, not lengthwise.

The holes for the wooden plugs can be cut with an ordinary countersink. Simply drill a bit deeper with the countersink than would normally be done for a regular countersunk screw. After the hole has been drilled, coat the plug with glue and insert it in place with the grain direction matching the wood itself. After the glue dries, the excess can be trimmed off with a chisel or sanded with abrasive paper. A piece of dowel rod can be used if the product is to be painted or if the plugs are to be part of the design, as in some Colonial furniture. The head of the screw can also be covered with wood putty.

Screwdrivers for Wood Screws

As mentioned, screws come in a variety of sizes, each of which has a slot of specific width and depth. To work well, the tip of the screwdriver should fit the slot as closely as possible. Since the slots vary in size, so must the screwdriver tips if you are to get a close fit. There are two basic types of screwdrivers; the standard or slotted-head screwdriver and the recessed-head screwdriver.

Screwdrivers for Slotted-Style Screws The so-called standard, or conventional, screwdriver is used for screws with slotted heads. They are classified according to the combined length of the shank and blade and by blade size. The most common sizes range in length from 2½ to 12 inches. The blade size varies from $3/32$ to ½ inch. However, the following

sizes will be a good match for average slotted-head screw sizes: $3/32$-inch tip with 3-inch shaft; ¼-inch tip with 4-inch shaft; $5/16$-inch tip with 6-inch shaft; and $3/8$-inch tip with 10-inch shaft. You also should have a stubby 1½-inch screwdriver with a ¼-inch tip for working in very tight places. There are, of course, many smaller and some larger screwdrivers for special purposes. The diameter of the shank and the width and thickness of the blade are generally proportional to the length, but again there are special screwdrivers with long thin shanks, short thick shanks, and extra-wide or extra-narrow blades. As a rule, however, blades and handles are proportioned in size to withstand any torque to which they are subjected in normal use.

Slotted Screw Tool Selector			
Screw Size	Blade Size, Inches	Screw Size	Blade Size, Inches
0	3/32	9	5/16
1	3/32	10	5/16
2	1/8	12	3/8
3	5/32	14	3/8
4	3/16	16	7/16
5	3/16	18	7/16
6	1/4	20	1/2
7	1/4	24	1/2
8	5/16		

For heavy work, special types of standard screwdrivers are made with a square shank which can withstand greater turning torque than round ones. Another advantage of a square shank is that you can put a wrench on it and supply some added torque when you have a rusted or frozen screw that is hard to turn. Some screwdrivers with round shafts have two flats on the shaft close to the handle for this purpose.

The blade of the standard screwdriver fans out just back of the tip. While this offers the convenience of a better grip on the screwhead when extra

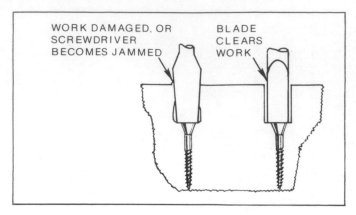

Standard blade and cabinet blade.

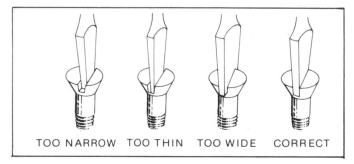

TOO NARROW   TOO THIN   TOO WIDE   CORRECT

Make sure the screwdriver blade fits the screw slot properly.

leverage is needed, it also prevents the tip from following a screw into its hole. When you wish to recess screws in deep holes, use a cabinet blade instead of the standard or keystone blade. The cabinet blade (also known as the gunsmith's blade) does not fan out; its tip is no wider than its shank.

When using a standard, or slotted, screwdriver, it is very important to select a screwdriver of the proper length whose tip is fitted properly to the screw. This proper fit is important because the tip tends to slip out of the slot if it is too thin, and it sometimes mars the work surface. More important, when the slot is too wide for the tip, the blade has room to twist in the slot and shave away fine slivers of steel from its edges. Since screws are made of relatively soft steel, the slot edges can be worn down in no time at all; then the screwdriver becomes useless because its blade has nothing to press against.

The width of the tip is as important as its thickness. If the tip is wide enough to protrude beyond the edges of the head of the screw, it will cut into the work surface as the screw is driven home. This is especially true when flathead screws are driven flush.

Screwdrivers for Screws with Recessed Openings The most common screw with a recessed slot is the Phillips screw. Its head has a four-way slot into which the screwdriver fits. This prevents

the screwdriver from slipping. Five standard-size Phillips screwdrivers handle a wide range of screw sizes. Their ability to hold helps to prevent damaging the slots or the work surrounding the screw. It is poor practice to try to use a standard screwdriver on a Phillips screw because both the tool and screw slot will be damaged. But when using a Phillips screwdriver, you must exert more downward pressure to keep the screwdriver in the slots. A worn Phillips screwdriver will tend to slip out of the slots and should be discarded, for it cannot be sharpened easily. As with the flat-bladed screwdriver, always use the correct size.

Phillips Screwdrivers		
Point Size	Blade Diameter Inches	Screw Gauge Number
0	1/8	0, 1
1	3/16	2, 3, 4
2	1/4	5, 6, 7, 8, 9
3	5/16	10, 12 14, 16
4	3/8	18, 20 24

Phillips head screwdrivers are sized by number from No. 0 (smallest) to No. 4. Generally, the sizes that the average craftsman is most likely to need are Nos. 2 and 3, with 4- and 6-inch shafts, respectively. A stubby 1½-inch Phillips screwdriver with a medium point (No. 2) is also useful for screws of this type in a tight place.

The other types of recessed-head screws must have their own special screwdrivers if the screwhead is to remain undamaged. The exception is Allen and hex-recess screws, for which both Allen and hex wrenches and Allen-bladed and hex-bladed screwdrivers may be used. Incidentally, screwdrivers having clutch-head tips lock tightly in the screwheads when turned clockwise. The screwdriver is unlocked simply by turning it in the opposite direction.

Tips on Driving Screws When driving any screw, remember these important points:

1. Use the blade which fits the screw slot exactly. Never use a round-tipped blade; it slips easily and may cause injury. Also, if the tip is rounded or beveled, it will rise out of the slot, spoiling the screwhead.

2. Use the longest screwdriver convenient for the job. More power can be applied to a long screwdriver than a short one, with less danger of its slipping out of the slot.

3. When starting a screw, place it on the screwdriver tip and hold the screw and tip together

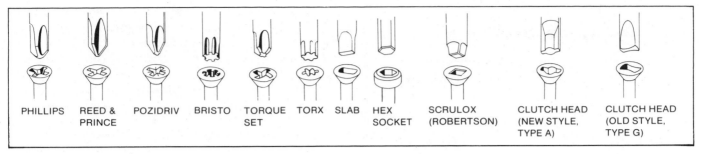

Recessed screws and screwdriver bits.

with the fingers of one hand, while grasping the handle with the other hand so it rests comfortably in the palm. Then put the screw point in the starting hole, retaining the same grip but permitting the hand holding the screw to rest on the work. The handle of a good screwdriver is rounded and smooth at the end (dome) to provide a comfortable palm rest, and the stopped flutes have rounded edges to minimize irritation and skin abrasion. A rubber crutch tip fitted over the handle of the screwdriver will make driving large screws easier and prevent palms from blistering.

4. To drive the screw in, turn the handle clockwise. To remove a screw, turn it counterclockwise. The screw should be at right angles to the surface unless the work requires that it be driven at an angle. Apply very little pressure on the driver at first, while turning it in a clockwise direction until the screw point engages the wood. As soon as the screw holds firmly, transfer your fingers to the screwdriver blade, letting it slip between them to keep the tip centrally on the screw shank. Apply just enough pressure on the driver to keep it in the slot, and hold the driver blade directly in line with the screw. Tilting the driver when starting a screw causes it to go in an angle, while tilting it after the screw is fully engaged may cause the driver to slip out of the slot. When cross-slot drivers are tilted, one of the wings is likely to break off because of unequal pressures.

5. If a screw is difficult to turn, back it out and enlarge the hole. Wax or paraffin rubbed on the screw will take most of the effort out of screwdriving. A candle is perfect for this purpose. Never use soap, because the moisture in the soap will rust the screw, weaken the joint, and cause discoloration of paint or varnish finishes. Do not use oil either; it will penetrate and stain the wood grain for some distance around the screwhead.

6. To speed up the driving of screws, there is sometimes a temptation to start them with a hammer. Do not do so. This pushes the wood fibers ahead of the screw and prevents it from taking a firm grip. It will not support its intended load.

7. Since a screw is often used with the idea that it may later be removed and replaced, be aware that, when withdrawn, it will leave a much larger hole than a nail will. Before the screw can be replaced, the hole must be filled to provide a firm gripping area. You can fill it with wood putty and reset the screw in this new material, or use plastic plugs driven into the old screw holes. Both provide an excellent base. You can also use a longer replacement screw.

8. To prevent flathead screws from working loose after they have been seated, try striking the head near the rim with a punch or a nail. The displaced metal serves as an anchoring wedge. Screws can also be driven by a power driver.

Other Types of Wood Screw Fasteners

The most popular of the other wood screw types are the lag screws. These screws (frequently called lag bolts) are often required in construction building. They are longer and much heavier than the common wood screw and have coarser threads which extend from a cone or gimlet point slightly more than half the length of the screw. Squarehead and hexagonhead lag screws are always externally driven, usually by means of a wrench. They are used when ordinary wood screws would be too short or too light and spikes would not be strong enough.

When using a lag screw, first drill a hole slightly larger than the diameter of the shank to a depth that is equal to the length that shank will penetrate. Then, drill a second hole at the bottom of the first hole equal to the root diameter of the threaded shank and to a depth of approximately one-half the length of the threaded portion. The exact size of this hole and its depth will, of course, depend on the kind of wood; the harder the wood, the larger the hole. To

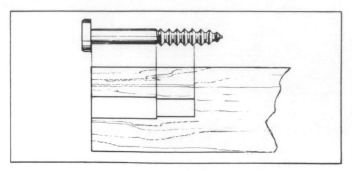

Drilling the hole for a lag screw.

Lag Screws				
Diameters (inches)				
Lengths (inches)	¼	⅜, 7/16, ½	⅝, ¾	⅞, 1
1	x	x	-	-
1½	x	x	x	-
2, 2½, 3, 3½, etc., 7½,				
8 to 10	x	x	x	x
11 to 12	-	x	x	x
13 to 16	-	-	x	x

make the turning of a lag screw easier, coat the threaded portion with paraffin wax.

The dowel screw is used for end-to-end joints and similar applications unsuited to conventional screws. The hanger screw (or bolt) has one end that is threaded like a screw so it can be driven into the wood; the other end is threaded to accept either a hex or square nut. Clove screws are frequently called

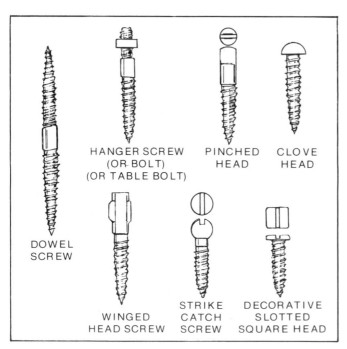

Other wood screw fasteners.

one-way screws since, once driven, they cannot be removed with a screwdriver; the slot is so designed that the blade will slip out of the slot. The basic idea of this screw is to prevent the theft of the objects they are holding.

There are several decorative-type screws on the market. The simplest of these is a cap that snaps over the screwhead.

Wood Bolts

Wood bolts are used in work when great strength is required or where the pieces being held must be frequently disassembled. Their use usually implies the employment of nuts for fastening and sometimes the use of washers to protect the surface of the material they are used to fasten. Bolts are selected for application to specific requirements in terms of length, diameter, threads, style of head, and type. Proper selection of head style and type of bolt will result in good appearance as well as good construction. The use of washers between the nut and a wood surface or between both the nut and the head and their opposing surfaces will prevent marring the surfaces and permit additional torque in tightening. The three most common wood bolts used by homecraftsmen are: carriage bolts, machine bolts, and stove bolts.

Carriage Bolts   Carriage bolts fall into three categories: square neck bolt, finned neck bolt, and ribbed neck bolt. These bolts have roundheads that are not designed to be driven. They are threaded only part of the way up the shaft; usually the threads are two to four times the diameter of the bolt in length. In each type of carriage bolt, the upper part of the shank, immediately below the head, is designed to grip the material into which the bolt is inserted and keep the bolt from turning when a nut is tightened down on it or removed. The finned type is designed with two or more fins extending from the head to the shank. The ribbed type is designed with longitudinal ribs, splines, or serrations on all or part of a shoulder located immediately beneath the head.

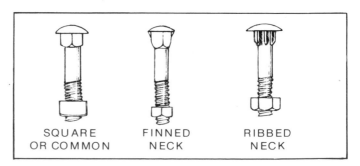

Carriage bolts.

Holes bored to receive carriage bolts are bored to be a tight fit for the body of the bolt and counterbored to permit the head of the bolt to fit flush with, or below the surface of, the material being fastened. The bolt is then driven through the hole with a hammer. Carriage bolts are chiefly for wood-to-wood application but may also be used for wood-to-metal applications. If used for wood-to-metal application, the head should be fitted to the wood item. Metal surfaces are sometimes predrilled and countersunk to permit the use of carriage bolts metal-to-metal. Carriage bolts can be obtained from ¼ inch to 1 inch in diameter, and from ¾ inch to 20 inches long. A common flat washer should be used with carriage bolts between the nut and the wood surface.

Carriage Bolts			
Diameters (inches)			
Lengths (inches)	3/16, 1/4 5/16, 3/8	7/16, 1/2	9/16, 5/8	3/4
¾	X	-	-	-
1	X	X	-	-
1¼	X	X	X	-
1½, 2, 2½, etc., 9½, 10 to 20	X	X	X	X

Machine Bolts These bolts are made with cut National Fine or National Coarse threads extending in length from twice the diameter of the bolt plus ¼ inch (for bolts less than 6 inches in length), to twice the diameter of the bolt plus ½ inch (for bolts over 6 inches in length). They are precision made and generally applied metal-to-metal where close tolerance is desirable. The head may be square, hexagon, rounded, or flat countersunk. The nut usually corresponds in shape to the head of the bolt with which it is used. Machine bolts are externally driven only. Selection of the proper machine bolt is made on the basis of head style, length, diameter, number of threads per inch, and coarseness of thread. The hole through which the bolt is to pass is bored to the same diameter as the bolt. Machine bolts are made in diameters from ¼ inch to 3 inches and may be obtained in any length desired.

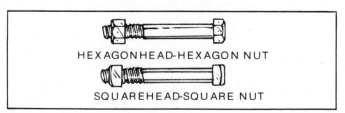

Machine bolts.

Machine Bolts				
Diameters (inches)				
Lengths (inches)	¼, 3/8	7/16	½, 9/16, 5/8	¾, 7/8, 1	1⅛, 1¼
¾	X	-	-	-	-
1, 1¼	X	X	X	-	-
1½, 2, 2½	X	X	X	X	-
3, 3½, 4, 4½, etc., 9½, 10 to 20	X	X	X	X	X
21 to 25	-	-	X	X	X
26 to 39	-	-	-	X	X

Stove Bolts These bolts are less precisely made than machine bolts. They are made with either flat or round slotted heads and may have threads extending over the full length of the body, over part of the body, or over most of the body. They are generally used with square nuts and applied metal-to-metal, wood-to-wood, or wood-to-metal. If flatheaded, they are countersunk; if roundheaded, they are drawn flush to the surface.

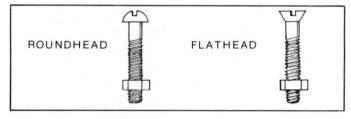

ROUNDHEAD   FLATHEAD

Stove bolts.

Tips on Fastening with Wood Bolts Keep the following points in mind when fastening wood pieces together with bolts:

1. Buy bolts long enough. The nut at the far end should go all the way onto the threaded portion of the bolt and leave a little over.
2. When drilling the hole through the wood, it should be exactly the same diameter as the bolt.

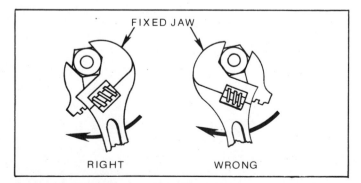

FIXED JAW

RIGHT   WRONG

Proper procedure for pulling adjustable wrenches.

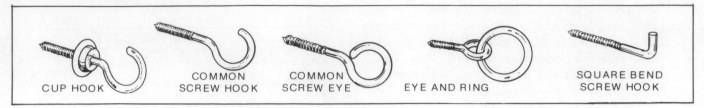

Wood screw hooks and eyes for supporting items on the wall.

3. Use washers both under the head and under the nut whenever possible.
4. When tightening the nut with a wrench, be sure the jaws of the wrench are the proper size for the nut. When using an adjustable wrench, pull the handle toward the side having the adjustable jaw. This will prevent the jaw from springing open and slipping off the nut.

Other Wood Bolts   There are two other types of bolts used in holding wood parts together or fastening wood items to other materials. These are driftpins and expansion bolts.

Driftpins are long, heavy, threadless bolts used to hold heavy pieces of lumber together. The term driftpin is almost universally used in practice. However, for supply purposes the correct designation is driftbolt. They have heads and vary in diameter from ½ to 1 inch and in length from 18 to 26 inches.

To use the driftpin, a hole slightly smaller than the diameter of the pin is made in the timber. The pin is driven into the hole and is held in place by the compression action of the wood fibers.

Wood Screw Hooks and Eyes

There are wood screw hooks and eyes that will either support various items on the wall or hold them together. For instance the common screw hook, which can be used to hang anything from kitchen utensils to shop tools, has a sharp point to make penetration into the wood easy and can be driven by hand or pliers to the depth required. The cup hook is fitted with a stop cap to make certain of the depth it will be driven, and it is frequently used for holding a cup in kitchen cabinets. The square bend hook is useful in supporting curtain rods or for hanging pots and pans.

While most common screw eyes are formed from a single piece of heavy wire, cast and forged types are available where greater strength is needed. Combined with a snap hook, the eye and ring can be used to hold doors and gates closed.

Screws and Bolts to Fasten Metal Parts

Metal parts can be fastened together with various fastening devices, such as rivets, bolts, screws, and so on. Rivets provide a more permanent type of fastening, whereas screws and bolts are used to fasten together parts that may have to be taken apart later.

Screws are the most common way of putting metal parts together. Like wood screws, there are a good number of types of metal screws. Let us take a look at the ones with which the home handyman will most likely come in contact.

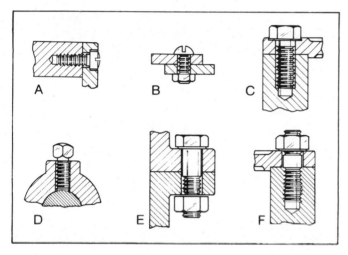

Common types of metal fasteners: (A and B) machine screws, (C) capscrew, (D) setscrew, (E) bolt, and (F) stud.

Machine Screws

The term machine screw is the general term used to designate the small screws that are used in tapped holes for the assembly of metal parts. Machine screws may also be used with nuts, but usually they are screwed into holes that have been tapped with matching threads.

Machine screws are manufactured in a variety of lengths, diameters, pitches (threads per inch), materials, head shapes, finishes, and thread fits. A complete description of machine screws must include these factors, for example "½-inch, 8-32, roundhead, brass, chromium-plated, machine screw." The first number is the length of the screw.

Diameter and Pitch The diameters of American Standard machine screws are expressed in gauge numbers or fractions of an inch. In the preceding paragraph, the 8-32 means that the screw gauge is Number 8 and that it has 32 threads per inch.

Materials and Finishes Most machine screws are made of steel or brass. They may be plated to help prevent corrosion. Other special machine screws made of aluminum or Monel metal are also obtainable. The latter metal is highly resistant to the corrosive action of salt water.

Head Shapes Except for the hexagon, the nine common heads are available in slotted form or with recesses. In addition, there are several special machine screwheads. Some of these heads require special tools for driving and removing. These special tools are usually included in a kit that comes with the machine or installation on which the screws are used.

Fits At one time each manufacturer made as many threads per inch on bolts, screws, and nuts as suited his own particular needs. For example, one made 12 threads per inch on ½-inch bolts while another put on 13 or 15 threads per inch. Thus, the bolts of one manufacturer would not fit the nuts made by another.

The National Screw Thread Commission studied the problem and decided to standardize on a two-thread series, one called the National Coarse Thread Series (NC) and the other the National Fine Thread Series (NF). The Society of Automotive Engineers decided to standardize on some Extra Fine (EF) threads to be used in airplanes, automobiles, and other places where extra fine threads are needed.

Four classes of fits were also established by the National Screw Thread Commission. They are: Class I, loose fit; Class II, free fit; Class III, medium fit; and Class IV, close fit.

The loose fit is for threaded parts that can be put together quickly and easily even when the threads are slightly bruised or dirty and when a considerable amount of shake or looseness is not objectionable. The free fit is for threaded parts that are to be put together nearly or entirely with the fingers and a little shake or looseness is not objectionable. This includes most of the screw thread work. The medium fit is for the higher grade of threaded parts where the fit is somewhat closer. The close fit is for the finest threaded work where very little shake or

Screw Threads per Inch				
Diameter	Threads per Inch			
No. or Fraction	Decimal Equivalent	NC	NF	EF
0	.0600	-	80	-
1	.0730	64	72	-
2	.0860	56	64	-
3	.0990	48	56	-
4	.1120	40	48	-
5	.1250	40	44	-
6	.1380	32	40	-
8	.1640	32	36	-
10	.1900	24	32	40
12	.2160	24	28	-
1/4	.2500	20	28	36
5/16	.3125	18	24	32
3/8	.3750	16	24	32
7/16	.4375	14	20	28
1/2	.5000	13	20	28
9/16	.5625	12	18	24
5/8	.6250	11	18	24
3/4	.7500	10	16	20
7/8	.8750	9	14	20
1	1.0000	8	14	20

looseness is desirable and where a screwdriver or wrench may be necessary to put the parts together. The manufacture of threaded parts belonging to this class requires the use of fine tools and gauges. This fit should, therefore, be used only when requirements are exacting or where special conditions require screws having a fine, snug fit.

Cutting Machine Screw Threads with a Tap
Machine screws are used in tapped holes for the assembly of metal parts. To cut these necessary internal threads, a tap is used.

There are three common taps: taper, plug, and bottoming. The taper (starting) hand tap has a chamfer length of 8 to 10 threads. These taps are used when starting a tapping operation and when tapping through holes. Plug hand taps have a chamfer length of 3 to 5 threads and are designed for use after the taper tap. Bottoming hand taps are used for threading the bottom of a blind hole. They

have a very short chamfer length of only 1 to 1½ threads, for this purpose. This tap is always used after the plug tap has already been used. Thus, both the taper and plug taps should be used before the bottoming hand tap.

A 50-50 mixture of white lead and lard oil, applied with a small brush, is highly recommended as a lubricant when tapping in steel. When using this lubricant, tighten the tap in the tap wrench and apply the lubricant to the tap. Start the tap carefully, with its axis on the center line of the hole. The tap must be square with the surface of the work.

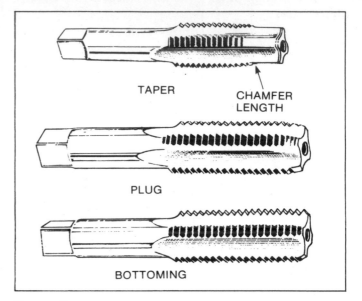

Types of common taps.

To continue tapping, turn the tap forward two quarter turns, back it up a quarter turn to break the chips, and then turn forward again to take up the slack. Continue this sequence until the threads are cut. After you have cut for the first two or three full turns, you no longer have to exert downward pressure on the wrench. You can tell by the feel that the tap is cutting as you turn it. Do not permit chips to clog the flutes or they will prevent the tap from turning. When the tap will not turn and you notice a springy feeling, stop trying immediately. Back up the tap a quarter turn to break the chips, clean

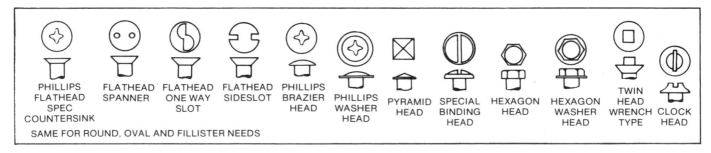

Special machine screw and capscrew heads.

American Standard Machine Screws

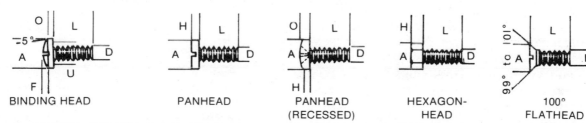

	Roundhead		Flathead	Fillister-head			Ovalhead		Trusshead	
Nominal diam.	A	H	A	A	H	O	A	C	A	H
0	0.113	0.053	0.119	0.096	0.045	0.059	0.119	0.021		
1	0.138	0.061	0.146	0.118	0.053	0.071	0.146	0.025	0.194	0.053
2	0.162	0.069	0.172	0.140	0.062	0.083	0.172	0.029	0.226	0.061
3	0.187	0.078	0.199	0.161	0.070	0.095	0.199	0.033	0.257	0.069
4	0.211	0.086	0.225	0.183	0.079	0.107	0.225	0.037	0.289	0.078
5	0.236	0.095	0.252	0.205	0.088	0.120	0.252	0.041	0.321	0.086
6	0.260	0.103	0.279	0.226	0.096	0.132	0.279	0.045	0.352	0.094
8	0.309	0.120	0.332	0.270	0.113	0.156	0.332	0.052	0.384	0.102
10	0.359	0.137	0.385	0.313	0.130	0.180	0.385	0.060	0.448	0.118
12	0.408	0.153	0.438	0.357	0.148	0.205	0.438	0.068	0.511	0.134
¼	0.472	0.175	0.507	0.414	0.170	0.237	0.507	0.079	0.573	0.150
5/16	0.590	0.216	0.635	0.518	0.211	0.295	0.635	0.099	0.698	0.183
⅜	0.708	0.256	0.762	0.622	0.253	0.355	0.762	0.117	0.823	0.215
7/16	0.750	0.328	0.812	0.625	0.265	0.368	0.812	0.122	0.948	0.248
½	0.813	0.355	0.875	0.750	0.297	0.412	0.875	0.131	1.073	0.280
9/16	0.938	0.410	1.000	0.812	0.336	0.466	1.000	0.150	1.198	0.312
⅝	1.000	0.438	1.125	0.875	0.375	0.521	1.125	0.169	1.323	0.345
¾	1.250	0.547	1.375	1.000	0.441	0.612	1.375	0.206	1.573	0.410

	Binding Head				Panhead			Hexagonhead		100° Flathead
Nominal diam.	A	O	F	U	A	H	O	A	H	A
2	0.181	0.046	0.018	0.141	0.167	0.053	0.062	0.125	0.050	
3	0.208	0.054	0.022	0.162	0.193	0.060	0.071	0.187	0.055	
4	0.235	0.063	0.025	0.184	0.219	0.068	0.080	0.187	0.060	0.225
5	0.263	0.071	0.029	0.205	0.245	0.075	0.089	0.187	0.070	
6	0.290	0.080	0.032	0.226	0.270	0.082	0.097	0.250	0.080	0.279
8	0.344	0.097	0.039	0.269	0.322	0.096	0.115	0.250	0.110	0.332
10	0.399	0.114	0.045	0.312	0.373	0.110	0.133	0.312	0.120	0.385
12	0.454	0.130	0.052	0.354	0.425	0.125	0.151	0.312	0.155	
¼	0.513	0.153	0.061	0.410	0.492	0.144	0.175	0.375	0.190	0.507
5/16	0.641	0.193	0.077	0.513	0.615	0.178	0.218	0.500	0.230	0.635
⅜	0.769	0.234	0.094	0.615	0.740	0.212	0.261	0.562	0.295	0.762

Thread length: screws 2 in. long or less, thread entire length; screws over 2 in. long, thread length 1 = 1¾ in. Threads are coarse or fine series, class 2. Heads may be slotted or recessed as specified, excepting hexagon form, which is plain or may be slotted if so specified. Slot and recess proportions vary with size of fastener; draw to look well.

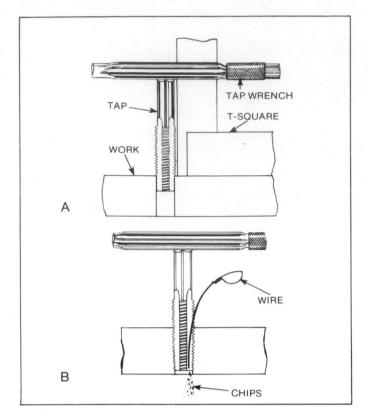

(A) Using a square to ascertain that a tap is square with the work. (B) Using a wire to clear chips from the flute of a tap.

them out of the flutes with a wire, add more lubricant, and continue tapping. When the tap has cut threads through the hole, the tap will turn with no resistance.

To tap a blind hole, start with the taper tap. For a blind hole you will need all three types—the taper, plug, and bottoming taps. (Be sure they are of the size and thread series you need.) Begin with the taper tap. Handle it as previously described.

The plug tap enters the full threads cut by the taper tap to continue these threads a little farther down into the hole. Finally, the plug tap is bottomed in the hole. This is all the work that you can do with this tap. It has cut full threads about halfway down the taphole before bottoming.

At this point, the bottoming tap is substituted for the plug tap. It runs down the full threads cut by the plug tap and is used to cut more full threads. Finally it is bottomed in the hole and the blind hole is completely tapped.

Because these threads are being tapped in a blind hole, the chips must be removed differently. To remove the chips, back the tap completely out of the hole very frequently, invert the stock, if possible, and jar out the chips or work them out of the hole with a wire while the stock is in the inverted position. Until these chips are removed, none of the

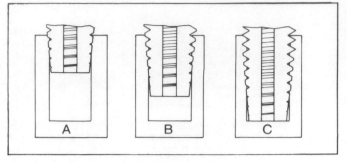

Tapping a blind hole with a taper tap.

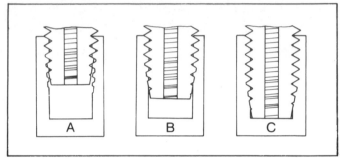

Tapping a blind hole with a plug tap.

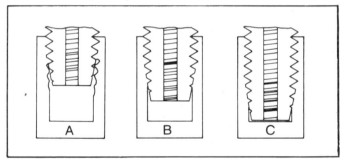

Finish tapping a blind hole with a bottoming tap.

three taps can complete its work. In tapping blind holes, alternate tapping and chip removal until each of the three taps bottoms in the blind hole.

When you have finished using the three taps, brush the chips out of their teeth, oil them well with lubricating oil, wipe off the surplus oil, and replace them in the threading set.

Machine screws are driven with a screwdriver in the same manner as wood screws.

Capscrews

Capscrews perform the same functions as machine screws, but come in larger sizes for heavier work. Sizes range up to 1 inch in diameter and 6 inches in length.

Capscrews are usually used without nuts. They pass through a clearance hole in one piece and

Tap Drill Sizes				
No. or Frac-tion	Tap		Tap Drill		Drill for Clearance
	NC	NF	NC	NF	
0		80		3/64	51
1	64	72	53	53	47
2	56	64	50	50	42
3	48	56	47	45	37
4	40	48	43	42	31
5	40	44	38	37	29
6	32	40	36	33	26
8	32	36	29	29	17
10	24	32	25	21	8
12	24	28	16	14	1
1/4	20	28	7	3	Same as tap
5/16	18	24	F	I	Same as tap
3/8	16	24	5/16	Q	Same as tap
7/16	14	20	U	25/64	Same as tap
1/2	13	20	27/64	29/64	Same as tap
9/16	12	18	31/64	33/64	Same as tap
5/8	11	18	17/32	37/64	Same as tap
3/4	10	16	21/32	11/16	Same as tap
7/8	9	14	49/64	13/16	Same as tap
1"	8	14	7/8	15/16	Same as tap

made of alloy steel and can withstand great stresses, strains and shearing forces.

Some capscrews have small holes through their heads. A wire, called a safety wire, is run through the holes of several capscrews to keep them from coming loose.

Setscrews

Setscrews screw into a tapped hole in an outer part, often a hub, and bear with their points against an inner part, usually a shaft. Setscrews are made of hardened steel and hold two parts in relative position by having the point set against the inner part. They are classified by diameter, thread, head shape, and point shape. The point shape is important because it determines the holding qualities of the setscrew.

Setscrews hold best if they have either a cone point or a dog point. These points fit into matching recesses in the shaft against which they bear.

Headless setscrews—slotted and socket-head such as Allen or Bristol types—are used with moving parts because they do not stick up above the surface. They are threaded all the way from point to head. Common setscrews, used on fixed parts, have square heads. They have threads all the way from the point to the shoulder of the head.

Thumb screws are used for setscrews, adjusting screws, and clamping screws. Because of their design, they can be loosened or tightened without the use of tools.

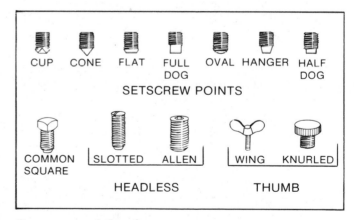

Setscrews and thumb screws.

screw into a tapped hole in the other. The head, an integral part of the screw, draws the parts together as the screw enters the tapped hole. The threads may be either NF or NC.

American Standard capscrews may have square, flat socket, round, or fillister heads. They are well-finished products; for example, the heads of the slotted- and socket-head screws are machined, and all have chamfered points.

Fillister-heads are best for use on moving parts when such heads are sunk into counterbored holes. Hexagonheads are usually used where the metal parts do not move. The strongest capscrews are

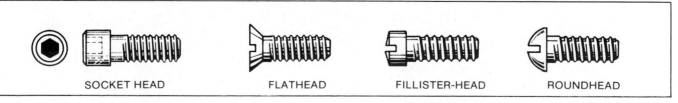

American Standard Capscrews.

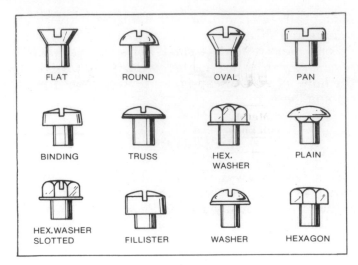

Typical self-tapping screwhead styles.

Self-Tapping Sheet Metal Screws

In addition to holding metal parts together, some self-tapping screws are ideal for joining sheet metal to wood and fastening together such man-made materials as hardboard, particleboard, and some plastics.

The most popular self-tapping screwhead styles are the pan-, flat-, round-, and fillister-head styles. These heads are available in both straight-slots and recessed-slots.

There are also five common types of self-tapping metal screws: A, B, C, F, and U. Their characteristics differ as follows:

• Type A have very coarse threads and are intended for joining sheet metal from Number 18 (0.048 inch) to Number 28 (0.015 inch) gauge. They can also be used to join metal to wood and to fasten plywood, particleboard, hardboard, and some plastic materials. These screws are commonly available in diameters of Number 4 through Number 14 and in lengths of ¼ to 2 inches. Type AB have the same characteristics but have finer threads.

• Type B have blunt points and finer threads than Type A. They are recommended for thicker gauges (Number 5 to Number 28), die casting, and so on. These screws are not used in woodworking. They are generally available in the same diameters and lengths as Type A.

• Type C have threads that approximate those of machine screws. They are used primarily for joining sheet metal that has a gauge which ranges from 12 to 22. These screws are usually available in lengths of about ¼ to 2 inches and diameters of Number 6 to ⅜ inches. Since Type C

screws do not cut chips but force the thread into the metal, they usually require considerable driving torque.

• Type F screws have, in effect, their own tap to cut threads as they are driven, like those of machine screw threads. They work well in aluminum, steel, and plastic in thicknesses from 0.050 to ½ inch. They are available in diameters of Number 2 to ⅜ inches and lengths from ¼ to 1½ inches.

• Type U screws are often called drive screws because they are installed with a hammer rather than a screwdriver. They are used to permanently attach nameplates and relatively thin materials (0.050 to ½ inch) to heavier metals,

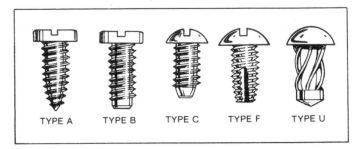

Types of self-tapping metal screws.

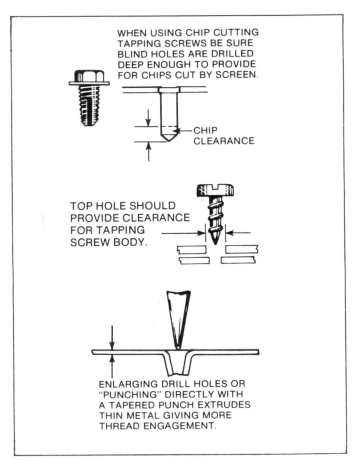

WHEN USING CHIP CUTTING TAPPING SCREWS BE SURE BLIND HOLES ARE DRILLED DEEP ENOUGH TO PROVIDE FOR CHIPS CUT BY SCREEN.

CHIP CLEARANCE

TOP HOLE SHOULD PROVIDE CLEARANCE FOR TAPPING SCREW BODY.

ENLARGING DRILL HOLES OR "PUNCHING" DIRECTLY WITH A TAPERED PUNCH EXTRUDES THIN METAL GIVING MORE THREAD ENGAGEMENT.

Self-tapping screwdriving tips to remember.

Suggested Metal Screw Drill Sizes

 TYPE A   TYPE B   TYPE U

Screw Size	Metal Thickness		Type A-AB Sharp Point
	Gauge	Inches	Use Drill
No. 4 (.112")	28	.016	No. 44
	26	.019	44
	24	.025	42
	22	.031	42
	20	.038	40
No. 6 (.138")	28	.016	No. 39
	26	.019	39
	24	.025	39
	22	.031	38
	20	.038	36
No. 7 (.155")	28	.016	No. 37
	26	.019	37
	24	.025	35
	22	.031	33
	20	.038	32
	18	.050	31
No. 8 (.165")	26	.019	No. 33
	24	.025	33
	22	.031	32
	20	.038	31
	18	.050	30
No. 10 (.191")	26	.019	No. 30
	24	.025	30
	22	.031	30
	20	.038	29
	18	.050	25
No. 12 (.218")	24	.025	No. 26
	22	.031	25
	20	.038	24
	18	.050	22
No. 14 (.251")	24	.025	No. 15
	22	.031	12
	20	.038	11
	18	.050	9

Screw Size	Metal Thickness		Type B Blunt Point	Size	Hole Size
	Gauge	Inches	Use Drill		
No. 4 (.112")	28	.016	No. 44	#6	1/8"
	26	.019	44		
	24	.025	43		
	22	.031	42		
	20	.038	42		
No. 6 (.137")	28	.016	No. 37	#8	9/64"
	26	.019	37		
	24	.025	36		
	22	.031	36		
	20	.038	35		
No. 7 (.151")	26	.019	No. 32	#10	5/32"
	24	.025	32		
	22	.031	32		
	20	.038	32		
	18	.050	31		
	16	.063	30		
No. 8 (.163")	26	.019	No. 32	#12	3/16"
	24	.025	32		
	22	.031	32		
	20	.038	32		
	18	.050	30		
No. 10 (.186")	26	.019	No. 27	#14	7/32"
	24	.025	27		
	22	.031	27		
	20	.038	27		
	18	.050	27		
No. 12 (.212")	24	.025	No. 19	5/16"	19/64"
	22	.031	19		
	20	.038	19		
	18	.050	18		
1/4 (.243")	22	.031	No. 13	3/8"	23/64"
	20	.038	13		
	18	.050	11		
	16	.063	8		

casting, and so on. They come in diameters of Number 2 to 3/8 inches and lengths of 1/8 inch through 3/4 inch.

There are a great many so-called special types of tapping screws—some that drill their own holes, some with inbuilt washers and locking devices, and some for use with special materials. For example, Type 21 are generally used to fasten leather, fabric, and cardboard to metal. They are available in diameter sizes of Number 6 to 3/8 inches and lengths from about 11/32 to 1/2 inch. These drive screws are installed with a hammer.

When installing self-tapping metal screws, it is necessary to drill a pilot. The top hole through the

 TYPE C   TYPE F

Screw Size	Material Thickness (Steel)					
	20 Ga.	16 Ga.	10 Ga.	1/8"	1/4"	1/2"
6—32	#33	#32	1/8"	#32	1/8"	-
8—32	#29	#26	#26	#27	#25	-
10—32	11/64	11/64	#16	11/64	#16	-
12—24	3/16	#10	13/64	3/16	#8	13/64
1/4—20	7/32	#1	15/64	7/32	#1	#1
5/16—18	L	L	19/64	J	L	L
3/8—16	-	-	-	11/32	23/64	23/64

sheet metal should provide clearance for the tapping screw body. When using chip-cutting tapping screws (Types B and F), be sure that the blind holes are drilled deep enough to provide for the chips cut by the screw. In some putting-it-together tasks, it is a good idea to enlarge the drilled holes or to punch directly with a tapered punch. This action will help extrude thin metal, giving more thread engagement.

Self-tapping screws can be given a more finished appearance and greater bearing surface by employing countersunk or finishing washers under their heads. When using a thin-gauge piece of metal and soft backing material, such as wood or composition board, flat-, oval-, and fillister-head screws can be neatly countersunk with a blow from a blunt-pointed punch.

Bolts, Studs, and Nuts

Bolts, studs, and nuts are very important metal fasteners. They are especially handy for putting together parts that may have to be taken apart.

Bolts   A bolt, having an integral head on one end and a thread on the other end, is passed through clearance holes in two parts and draws them together by means of a nut screwed on the threaded end.

Two major groups of bolts have been standardized: roundhead bolts and machine bolts (sometimes called wrench-head bolts).

Roundhead bolts are used as through fasteners with a nut, usually square or hexagonal. Eleven head types have standard proportions and include carriage bolts, step bolts, spline bolts, and stove bolts. Several head types are intended for wood construction, and have square sections, ribs, or fins under the head to prevent the bolts from turning. These bolts are hot or cold formed, with no machin-

ing except threading; hence they present a somewhat coarse and irregular appearance.

Machine (wrench-head) bolts, have two standard head forms: square and hexagonal. Although intended for use as a through fastener with a nut, machine bolts are sometimes used as capscrews. Nuts to match the bolthead form and grade are available, but any nut of correct thread will fit a bolt. Machine bolts vary in grade from coarsely finished products resembling roundhead bolts to a well-finished product matching a hexagon capscrew in appearance.

Whenever possible, install bolts in metal parts so that their heads are up. This way, if the nut has been improperly secured or is shaken loose by vibration and falls off, the bolt will remain within the part and continue to retain its holding capability although the nut is missing.

Be certain that the grip length of the bolt is correct. The grip length is the length of the unthreaded portion of the bolt shank. Generally speaking, the grip length should equal the thickness of the material which is being bolted together. Not more than one thread should bear on the material, and the threaded portion of the shank should be showing the nut. If no threads from the bolt shank show through the nut extremity, too many threads are bearing on the material. The nut cannot be threaded far enough to apply pressure on the material.

Proper selection of bolt material is very important. Be sure when you are required to replace bolts (or any type of fastener) that you select the type of material designed for the equipment. Remember that dissimilar materials can cause erosion and galvanic action that could result in bolt failure.

Studs   A stud is a rod threaded on each end. As used normally, the fastener passes through a clearance hole in one piece and screws permanently into a tapped hole in the other. A nut then draws the

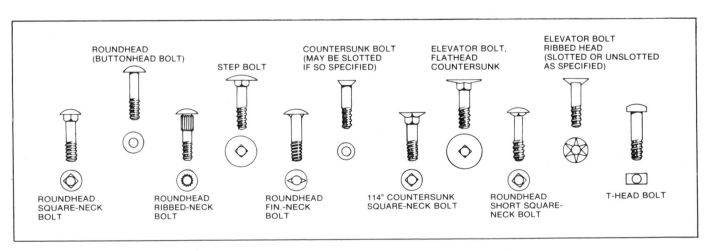

Some of the more popular types of bolts.

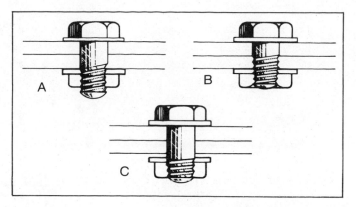

Correct (A) and incorrect (B and C) grip lengths.

screw at A. Tightening this screw spreads the die slightly so it will cut less deeply into the rod, and the fit in the tapped hole will be tighter. The shallow hole at B is placed in the diestock opposite the adjustable handle E and serves as a drive hole. Also, when the adjustable handle is tightened, it holds the split die together and against the adjusting screw to maintain the setting while the die is cutting. The threads or cutting teeth of the die are chamfered or relieved at C to help start the die squarely on the round stock. The die is put into the diestock so that the face with the unchamfered teeth is against the shoulder D.

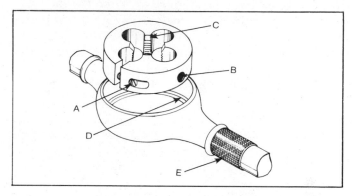

Assembling an adjustable round split die to the diestock.

parts together. The stud is primarily used when through bolts are not suitable for parts that must be removed frequently. One end is screwed tightly into a tapped hole, and the projecting stud guides the removable piece to position. The end to be screwed permanently into position is called the stud end, and the opposite end, the nut end. The nut end is sometimes identified by rounding instead of chamfering. Studs have not been standardized. The length of thread on the stud end is governed by the material tapped. The threads should jam at the top of the hole to prevent the stud from turning out when the nut is removed. The fit of the thread between the stud and tapped hole should be tight. The length of thread on the nut end should be such that there is no danger of the nut binding before the parts are drawn together. The name stud bolt is often applied to a stud used as a through fastener with a nut on each end.

Although studs are not standardized, they can be made from ready-made threaded rods or cut with a die. The threaded rods are available in various National Coarse threads and in diameter sizes from 3/16 to 2 inches. Their lengths range from 1 to 6 feet, and they can easily be cut to any stud size desired with a hacksaw.

Cutting Threads with a Die Once you have determined the proper diameter of round rod stock needed, grind a chamfer on the end of the rod. Then hold the rod vertically in a vise to cut the threads. The adjustable round split die has an adjusting

In assembling a plain round split die to the diestock, at A, where the die is split, there is no adjusting screw. There are shallow holes at B and C, on both sides of the split, opposite which there are setscrews in the diestock at D and E. F is the adjusting screw, which is pointed, and which enters the split A in the die. D and E are the holding setscrews.

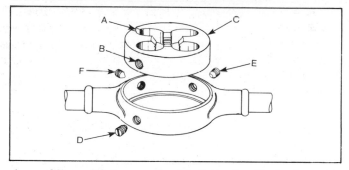

Assembling a plain round split die to the diestock.

Bolt-Length Increments										
Bolt diameter		1/4	5/16	3/8	7/16	1/2	5/8	3/4	7/8	1
Length increments	1/4	3/4—3	3/4—4	3/4—6	1—3	1—6	1—6	1—6	1—4 1/2	
	1/2	3—4	4—5	6—9	3—6	6—13	6—10	6—15	4 1/2—6	3—6
	1	4—5	-	9—12	6—8	13—24	10—22	15—24	6—20	6—12
	2	-	-	-	-	-	22—30	24—30	20—30	12—30

Example: 1/4" bolt lengths increase by 1/4" increments from 3/4 to 3" in length. 1/2" bolt lengths increase by 1/2" increments from 6 to 13" length. 1" bolt lengths increase by 2" increments from 12 to 30" length.

They have flat points and are tightened after the setting is made with F. D and E hold the adjustment and furnish the drive as they enter the shallow holes B and C.

Notice a section of the die in the diestock and its relation to the chamfer on the end of the work. The taper on the face of the die will accept the chamfer on the end of the work to start the threads square with the common center line.

To thread the work, brush some 50-50 white lead and lard oil on the rod. Start the die square with the work. Hold one handle with each hand, apply downward pressure, and turn clockwise until you feel that the thread has been started. When the die has started to cut, rotate the diestock two quarter turns, back it off one quarter turn to break the chips, and repeat the cutting. When you have cut enough threads so that the rod comes through the back of the die, remove the die and try the rod in the tapered hole.

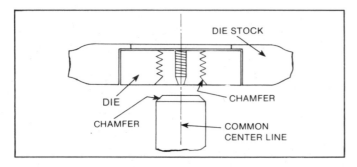

Position of a diestock in relation to the chamfer on the end of the work.

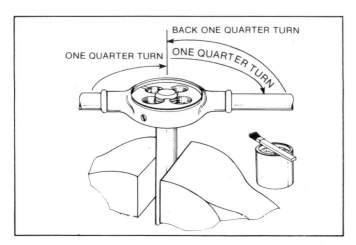

Cutting outside the threads on round rod stock.

Nuts   Like screws and bolts, there are many different types of nuts. Square and hexagonal nuts are standard, but they are supplemented by special nuts. One of these is the jam nut, used above a standard hexagon nut to lock it in position. It is about

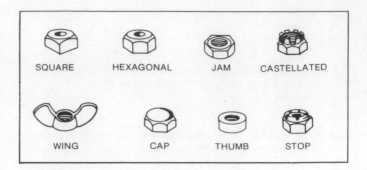

Common kinds of nuts.

half as thick as the standard hexagon nut and has a washer face.

Castellated nuts are slotted so that a safety wire or cotter key may be pushed through the slots and into a matching hole in the bolt. This provides a positive method of preventing the nut from working loose. For example, you will see these nuts used with the bolts that hold the two halves of an engine connecting rod together.

Wing nuts are used where the desired degree of tightness can be obtained by the fingers. Cap nuts are used where appearance is an important consideration. They are usually made of chromium-plated brass. Thumb nuts are knurled, so they can be turned by hand for easy assembly and disassembly.

Elastic stop nuts are used where it is imperative that the nut does not come loose. These nuts have a fiber or composition washer built into them which is compressed automatically against the screw threads to provide holding tension.

While a wrench—open-end, box, adjustable, or socket—is generally used for tightening hexagon and square nuts, smaller sizes can be handled by a nut driver. This driver is a screwdriver-type tool which, in its simplest form, has a one-piece shank and socket secured in a fixed handle. The socket heads have openings for hexagon and square nuts, boltheads, and screwheads up to ¾-inch nut size.

Washers   Flat washers are used to back up boltheads and nuts, and to provide larger bearing surfaces. They prevent damage to the surfaces of the metal parts.

Split lock washers are used under nuts to prevent loosening by vibration. The ends of these springhardened washers dig into both the nut and the work to prevent slippage. Shakeproof lock washers

Basic types of washers.

have teeth or lugs that grip both the work and the nut. Several patented designs, shapes, and sizes are obtainable.

Most of the nuts and washers described here can be used for wood fastening operations.

Keys and Pins   Various keys and pins are used by equipment manufacturers to retain parts in their proper position or to preserve their alignment. You should be able to identify these keys and pins, order replacements, and then replace them. Instructions for the latter are generally given in the equipment's service manual. Here are some of the more popular keys and pins.

Square keys and Woodruff keys are used to prevent handwheels, gears, cams and pulleys from turning on a shaft. These keys are strong enough to carry heavy loads if they are fitted and seated properly.

Round taper pins are used to locate and position matching parts. They are also used to secure small pulleys and gears to shafts. They usually have a taper of ¼ inch per foot. Holes for taper pins must be reamed with tapered reamers. If this is not done, the taper pin will not fit properly.

Dowel pins are used to position and align the units or parts of an assembly. One end of a dowel pin is chamfered, and it is usually .001 to .002 inch greater in diameter than the size of the hole into which the pin will be driven. When it is necessary to replace a dowel pin, be sure that it is the same size as the old one.

Cotter pins are used to secure screws, nuts, bolts, and pins. They are also used as stops and holders on shafts and rods. Some cotter pins are made of low-carbon steel, while others consist of stainless steel and thus are more resistant to corrosion. Regardless of shape or material, all cotter pins are used for the same general purpose—safetying.

The cotter pin should fit neatly into the hole with very little sideplay. In the preferred installation method, the bent prong above the bolt end should

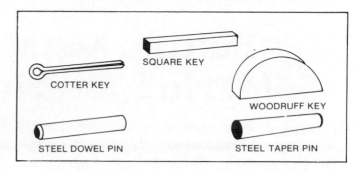

Keys and pins.

not extend beyond the bolt diameter. Additionally, the bent prong should not rest against the surface of the washer. Cut the prongs down to size if necessary.

If the optional wraparound method is used, the prongs should not extend outwards but should be bent over a reasonable radius to the sides of the nut. Sharply angled bends invite breakage. Usually the initial bending of the prongs of a cotter pin is accomplished with needle nose or diagonal pliers and the best tool for final bending of the prongs is a soft-faced mallet.

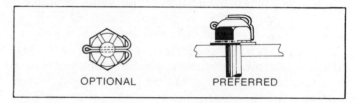

Cotter pin installation.

Other Threaded Fasteners

There are many other forms of threaded fasteners on the market that serve special purposes. For instance, the turnbuckle is an excellent tightening device. Consisting of a sleeve and two screw eyes (one with a right-hand thread, the other with a left), the turnbuckle is tightened or loosened by turning the sleeve.

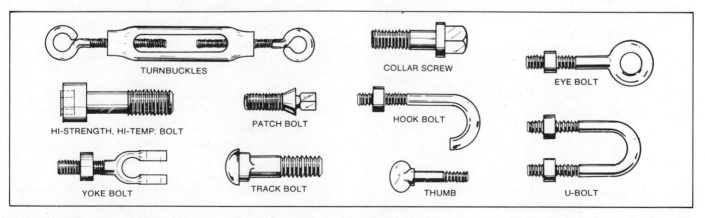

Various other threaded fasteners that serve special purposes.

Putting Metal Together by Soldering, Brazing, and Welding

In addition to the methods of putting metal together already described, there are three other popular metal-joining processes: soldering, brazing, and welding.

Soldering is done when a low-melting filler alloy (called solder) is heated to the point where it melts and is placed so that it wets the joint surfaces. It is then allowed to solidify in place. Actually, soldering is a practical method of forming reliable electrical connections where bare wires are twisted together or are wound on terminals. Soldering is also used to make tight joints, such as lap seams in sheet metal, and to hold parts together physically. Soldered joints, however, do not support loads for long periods of time as well as welded joints do. Where load support is a governing factor, the usual practice calls for riveting and bolting (as discussed earlier in this book), or for brazing and welding.

Brazing, or hard soldering as it is sometimes called, is the joining of metals by the fusion of nonferrous alloys that have melting points above 800 degrees F. but lower than those of the metals being joined. This may be done with a torch (torch brazing), in a furnace (furnace brazing), or by dipping in a molten flux bath (dip or flux brazing). The filler metal is ordinarily in rod form in torch brazing. However, in furnace and dip brazing the work is first assembled, and the filler metal may then be applied as wire, washers, clips, bands or it may be integrally bonded, as in torch brazing. In this book, only torch brazing will be covered.

Welding is a process used to join metals by the application of heat in which the work and the filler (if used) are actually melted so that they flow together and are integrally joined when cooled. Theoretically, a good weld can be made by melting the two metals together, but in actual practice a filler or a welding rod is used to fill the gaps and to smooth the finished joint. Thus, both the workpiece and the filler are of the same material—such as cast iron, steel, or aluminum.

Soldering

To solder the readily solderable metals, you need only the solder, a cleaning substance frequently called flux, and a heat source. Copper, tin, lead, and brass are examples of readily solderable metals. Galvanized iron, stainless steel, and aluminum are difficult to solder and require the use of special techniques which are discussed later in this section.

Solders By definition, solders are joining materials or alloys that melt below 800 degrees F. Often called soft solders, they are available in various forms: wire, bar, ingot, paste, and powder. The solders used for electrical connections are alloys of tin and lead whose melting points range between 360 and 465 degrees F. (both extremes are approximate).

A tin-lead solder alloy is usually identified by two numbers that indicate the percentages of tin and lead in the alloy. The first number is the percentage of tin. Thus, a 60-40 alloy is made of 60 percent tin and 40 percent lead. Likewise, a 40-60 alloy is made of 40 percent tin and 60 percent lead. In general, the higher the percentage of tin in a solder alloy, the lower the melting point. Aluminum, stainless steel, and a few other metals require special solders.

Fluxes Soldering fluxes are agents which clean solderable metals by removing the oxide film normally present on the metals and prevent further oxidation. Fluxes, classified as noncorrosive, mildly corrosive, or corrosive, range from mild substances such as rosin to chemically active salts such as zinc chloride. Rosin is an effective and nearly harmless flux used for electrical connections that must be reliable, tight, and corrosion-free. Rosin flux is available either in paste or powder form for direct application to joints before soldering or incorporated as the core of wire solders. Unless washed off thoroughly after soldering, salt-type fluxes leave residues that tend to corrode metals. Because of their corrosive effects, acid-core solders (which incorporate salt-type fluxes) must not be used in soldering electrical connections.

Solder is manufactured with a flux core and in a solid type. The former is most popular with hobbyists and do-it-yourselfers. Liquid or paste flux can be brushed on a joint when necessary. For instance, you will need flux if you are soldering with solid or bar solder which does not contain a core of flux. Also, if the solder is allowed to remain on the tip of the iron or gun for any period of time, the flux boils out of the core and must be replaced.

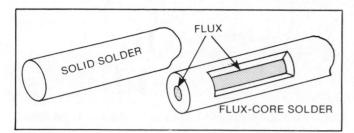

Two basic types of solder.

Heat Source The source of heat for melting solder is a soldering gun (electric) or a soldering iron (electric or nonelectric), sometimes called a copper.

Soldering Guns Most soldering guns use any standard 120-volt outlet and are rated by the number of watts they consume. The guns in general use are rated between 100 and 250 watts. All quality soldering guns operate in a temperature range of 500 to 700 degrees F. The important difference in gun sizes is not the temperature, but the capacity of the gun to generate and maintain a satisfactory soldering temperature while giving up heat to the joint being soldered. The tip heats only when the trigger is depressed, and then very rapidly. Some guns feature two heating temperatures that are usually controlled by the trigger. For example, the low temperature can be attained when the trigger is depressed halfway and the high when it is fully depressed.

Soldering guns afford access to cramped quarters, because of their small tips. Most soldering guns have a small light that is focused on the tip working area. Small cordless soldering irons are also available, as well as so-called field irons which can be operated from suitable batteries or low-voltage power sources supplying 12 to 14 volts.

Many soldering guns come with three tips designed for different operations. For soldering, be sure to use the soldering tip. The other two are for

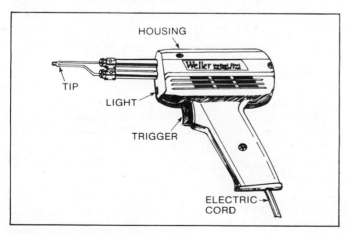

Parts of a typical soldering gun.

cutting and smoothing plastic materials. Incidentally, the latter is ideal for putting together plastic toys that literally come apart at the seams. To aid in making these repairs, use a plastic filler such as an old knitting needle.

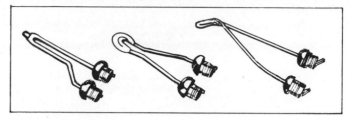

Three common types of tips: (Left to right) soldering, cutting, and smoothing.

Soldering Irons There are two general types of soldering irons in use. One is electrically heated, and the other is nonelectrically heated. The essential parts of both types are the tip and the handle. The tip is made of copper.

The nonelectric soldering iron is sized according to its weight. The commonly used sizes are the ¼-, ½-, ¾-, 1-, 1½-, 2-, and 2½-pound irons. The 3-, 4-, and 5-pound sizes are not used in ordinary work. Nonelectric irons have permanent tips and must be heated over flame, or with a blowtorch or propane torch.

The electric soldering iron transmits heat to the copper tip after the heat is produced by electric current which flows through a self-contained coil of resistance wire called the heating element. Electric soldering irons are rated according to the number of watts (from 25 to over 400 watts) they consume when operated at the rated voltage. A 50-watt electric iron is good for most small work, while a 100-watt iron is recommended for practically all home soldering. A 200-watt electric iron is suggested for heavier soldering, and for rugged work a 350-watt electric iron should be used.

Two types of tips are used on electric irons: plug tips which slip into the heater head and which are held in place by a setscrew, and screw tips which are threaded and which screw into or on the heater head. Some tips are offset and have a 90 degree angle for soldering joints that are difficult to reach.

Electric-iron tips must be securely fastened in the heater unit. The tips must be clean and free of copper oxide. Sometimes the shaft oxidizes and causes the tip to stick in place. Remove the tip occasionally, and scrape off the scale. If the shaft is clean, the tip will receive more heat from the heater element, and you will be able to remove the tip easily when it must be replaced.

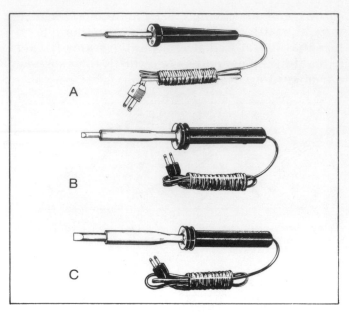

Typical electric soldering irons: (A) 25 watt, (B) 80 watt, and (C) 175 watt.

Soldering Procedures

A cardinal rule when doing any kind of soldering work is cleanliness. Be sure that all parts to be soldered are absolutely clean (free of oxide, corrosion, and grease). During the cleaning process, care must be taken not to produce cuts or nicks which greatly reduce the mechanical strength of the material, especially under conditions of vibration.

Do not touch the wire with your hands after it has been cleaned. Natural oils in the skin may cause the solder not to stick. Although flux is contained in the core of some solders, additional flux is sometimes required on extremely difficult jobs. For instance, if you find it difficult to get solder to stick on galvanized metal or any other extremely hard-to-solder surface, add some additional flux. This will normally improve the sticking capacity of the solder. Liquid flux can be brushed on the metal if required.

The joint should be prepared just prior to soldering since the prepared surfaces will soon corrode or become dirty if they remain exposed to the air. The parts to be soldered must be securely joined mechanically before any soldering is done. Remember that solder should be thought of as the final binding of a joint not as the primary means of strength. Any joint that will have a lot of stress applied to it is not a candidate for soldering.

Soldering Wires When putting wires together, the insulation should be removed carefully to expose each wire. Clean the bare wires with metal wool or fine emery cloth. If the wire is extremely dirty, it should be dipped into rosin flux—never acid flux.

Cross the bare wires and then twist them together to form a strong mechanical link.

To solder the wires, hold the hot soldering iron or gun beneath the splice being soldered with as much mechanical contact as possible to permit maximum heat transfer. Apply rosin-core solder to the splice. The tinning on the soldering iron aids in the transfer of heat to the spliced wires which, when hot enough, will melt the solder. Before this temperature is reached, the rosin core will have melted and run out over the wire to flux the splice. When the solder has coated the splice completely, the job is finished.

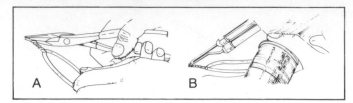

(A) Be sure to make a strong mechanical joint; (B) then, apply the heat to the wires, not the solder.

No extra solder is needed. A good, well-bonded connection is clean, shiny, smooth, and round. It also approximately outlines the wire and/or terminal. Let the soldered joint become completely cool before applying any pressure to it. After the solder has cooled and hardened, test it to make sure the soldered joint is secure.

Tinning wires is always a good practice before soldering them together. This is especially true with stranded wires. Twist the strands, then heat and saturate them with solder. Frequently, you will need

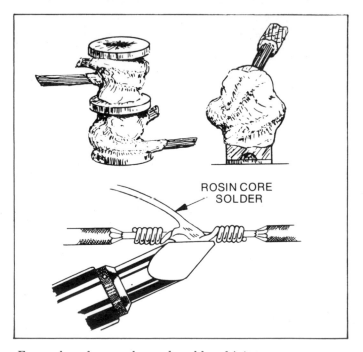

Examples of properly made soldered joints.

a small bench vise or some other holding device to provide a third hand when doing soldering jobs such as tinning. A small paper cup also makes an excellent holding device. A slot in each side of the cup will hold the wire in a firm position. It would be wise, however, to drop a small piece of asbestos into the bottom of the cup so the hot drippings from the soldering iron or gun will not cause a flame-up from the wax cup.

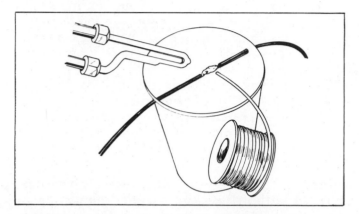

A paper cup makes a good holder when soldering wires.

Sweating a joint is easy after the wires are tinned. Just twist the two wires together and apply heat. The result will be a smooth, electrically-efficient soldered joint. Splices in the wires should be located at different positions. This eliminates the danger of shortages and lessens the amount of buildup when the soldered spots are taped for insulation. Cover the soldered area thoroughly with a high grade of insulation tape, or use shrinkable tubing to insulate the splice. Heat from the soldering tip makes the tubing shrink tight.

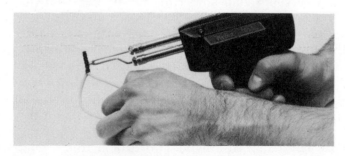

Applying shrinkable tubing.

When wiring a terminal lug, the back tabs are bent around the insulation, and wire is soldered to the terminal. Then, the front tabs are bent over the soldered wire. Add more solder to the front tabs where they touch the wire.

Board terminals for soldered connections usually have a hole, or eye. Bend a hook in the end of the wire, place it through the eye, and solder. Remem-

ber, a good, well-bonded connection approximates the outline of the joint.

Prolonged heat will weaken or break the wire and possibly damage nearby electronic components. Use normal heat and a pair of long-nose pliers as a heat sink to prevent overheating. As mentioned earlier, do not use more solder than needed; apply just enough to make a good connection. Excess solder may fill up the tube sockets, freeze the switches, or cause short circuits.

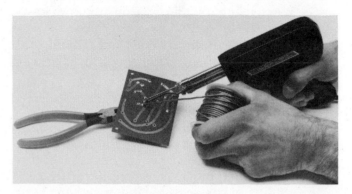

Using pliers as a heat sink.

Soldering Flat Pieces of Metal Most flat metals such as tin and copper should be soldered with a rosin-core solder. Acid-core solders should be used only on galvanized iron and other hard-to-solder metals. To get a good bond on two pieces of flat metal, a thin layer of solder should be applied to both edges. The tinned edges should be placed one over the other and pressed firmly in place with the hot tip of the gun or the broad side of the iron. As pressure is applied with the soldering tool, feed additional solder into the joint from the side. A little experience will enable you to sweat the edges and solder the two pieces of metal together easily, quickly, and firmly.

Slide the tip away while the tool is still hot. When using a gun, the trigger should never be released as the tip is being removed from the solder joint. Allowing the tip to cool while in contact with the solder will result in an untidy joint. Also, avoid any joint

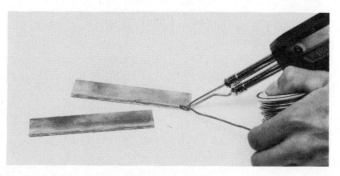

To get a good bond on two pieces of flat metal, a thin layer of solder should be applied to both edges.

movement. The joint must remain perfectly still until the solder sets. Pliers, a vise, or some other holding tool should be used. Also, blowing on the heated solder will speed the cooling period.

If you are attempting to solder any coated surface, such as enamelware, the coated area must be chipped away before the solder may be applied. Solder will not stick to coated surfaces.

Soldering Precautions One sizzling burn is usually enough to produce a healthy respect for hot objects. When using a soldering iron or gun, always bear in mind the following:

1. Electric soldering irons must not remain connected longer than necessary and must be kept away from flammable material.
2. To avoid burns, always assume that a soldering iron is hot.
3. Never rest a heated iron anywhere but on a metal surface or rack provided for this purpose. One uses two crossed finish nails tacked into a scrap piece of lumber. These nails keep the iron up off the flat surface, hold it in place, and keep the point of the iron clean while you are doing the job. An empty tin can also makes an ideal holder for a hot soldering iron. This holder can be mounted on a wall with a small piece of asbestos behind it to prevent damage to the wall from the heat of the iron.
4. Never swing an iron or gun to remove solder because the bits of solder that come off may cause serious skin or eye burns or may ignite combustible materials in the work area.
5. When cleaning an iron, use a cleaning cloth or damp sponge, but do not hold the cleaning cloth or damp sponge in your hand. Always place the cloth or damp sponge on a suitable

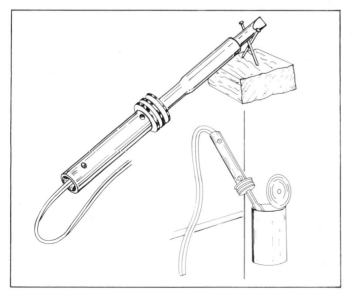

Two soldering-iron holders.

surface and wipe the iron or gun across it to keep from burning your hand.
6. Hold small soldering jobs with your pliers or a suitable clamping device. Never hold the work in your hand.
7. After completing a task requiring the use of a soldering iron, disconnect the power cord from the receptacle and, when the iron has cooled off, stow it in its assigned storage area. Do not throw irons into a toolbox. When storing a soldering iron or gun for a long period of time, coat the shaft and all metal parts with a rust-preventive compound and store it in a dry place.
8. When using a soldering gun, never tape back the trigger. Most guns are designed to provide heat intermittently, not constantly. An ideal work rate for most guns is 20 percent per minute (i.e., 12 seconds work followed by 48 seconds rest).

Care and Maintenance of a Gun A soldering gun needs little in the way of care and maintenance to keep it in top working condition. Occasionally, the soldering tip should be re-tinned when it gets dirty or when the solder, instead of flowing smoothly and clinging to the tip, rolls off in balls.

To re-tin the tip, depress the trigger and apply the solder until it begins to melt. Keep the tip in contact with the metal for three or four seconds, release the trigger, then continue to apply solder until the tip cools. Wipe off the excess solder and flux. If the tip is not thoroughly tinned, repeat the above procedure.

To clean the tip, rub it lightly with metal wool or fine emery cloth. (Never use a file.) After cleaning in this manner, the tip must be re-tinned.

When using your soldering gun be sure that the tip nuts are tight, as they can loosen during use. If the tool has not been used for some time, an oxide film may form at the tip nuts. To overcome this film, the nuts should be slackened and then retightened.

Care and Maintenance of a Soldering Iron Before soldering with any iron, be sure the tip is thoroughly cleaned. If corrosion is built up on the tip of the soldering point, it should be filed away with a light or medium file. The tip of an iron should be clean at all times. Cleaning the soldering iron after each use will eliminate much of the need for filing of the tip.

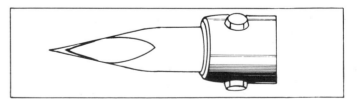

Proper shape of soldering-iron tip.

The shape of the tip of a soldering iron is also important. The modified chisel tip is ideal for most soldering jobs done with a soldering iron. The tip, of course, should be small enough to reach into tight places but blunt enough to insure that the heat will be transmitted all the way down to the point.

Before beginning the soldering job, a thin, even coat of solder should be applied to all sides of the iron. This coating process is frequently referred to as tinning.

You will find a regular kitchen-type cleaning pad or a piece of metal wool a handy cleaning device for keeping the point of the iron clean while you are soldering. This pad or piece of metal wool can be stapled or tacked to the work surface where you will be soldering. An occasional wipe across the cleaning pad will keep the point clean at all times.

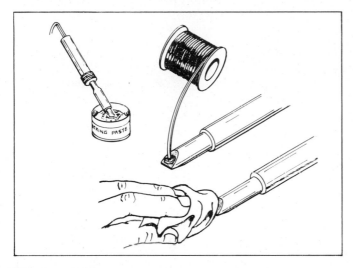

Tinning an iron.

Propane or Gas Torches

Propane or gas torches are frequently used in various soldering jobs around the home and in craft and hobby work. These torches operate on a very simple principle. The torch tank is filled with propane, a liquefied petroleum gas. The propane remains in a liquid state because it is under pressure. When the valve on the burner assembly is opened, propane is released through the fuel orifice. With reduced pressure, the propane converts to a gas, which burns readily when mixed with air. The tool consists of only two basic parts: the fuel tank and the burner assembly.

Some manufacturers have torches available that use Mapp gas rather than propane. A stabilized mixture of methylacetylene and propadiene, Mapp gas produces temperatures in excess of 3,200 degrees, or about 500 degrees hotter than propane.

Although slightly more expensive, it is frequently recommended for metal operations where the added heat is needed.

The fuel cylinder (tank) is constructed of heavy-gauge steel that meets rigorous safety specifications. It is engineered for adequate strength. Every tank is thoroughly tested for leaks before it leaves the factory. For easy handling, the tank is well proportioned and has a wide base that resists tipping. The gas dispensing valve at the top of the tank opens and closes automatically for convenient, economical torch assembly. When the burner assembly is screwed onto the tank, the valve releases propane if the gas-dispensing knob is opened. (Caution: Always inspect the inside of the cylinder outlet to make certain foreign matter has not become lodged in this area. Inspect before attaching the burner assembly.) When the burner head is removed from the tank, the valve closes automatically so that no propane is wasted. Every tank has an automatic pressure-release valve that eliminates the danger of the tank exploding. The valve opens automatically if the pressure in the tank is increased to a predetermined level which is above normal conditions but below unsafe conditions. Most Mapp-gas torches also have a fuel regulator. This eliminates the need for fuel adjustments since it automatically correlates fuel with flame size.

The burner-assembly unit screws onto the top of the propane or Mapp-gas tank. It can be tightened by hand; a wrench should not be used. The unit is produced under rigid quality-control standards and is thoroughly flame-tested at the factory. Caution should be exercised to prevent dropping material particles down into the center inlet valve area. Such particles could wedge the valve pin into the open position, preventing normal cylinder-valve closure upon separation of the cylinder and appliance. The burner assembly consists of a burner head, an air-intake hole, a fuel tube, a valve assembly, and a valve control knob. A variety of burner heads or tips are available and interchangeable, to permit you to select the right type of flame for every job.

Propane and Mapp-gas torches assemble quickly and easily. First, be sure the valve on the burner assembly is completely closed by turning the on/off control knob clockwise as far as it will go. Screw the burner head onto the tank (clockwise) until it is finger-tight. Avoid crossfeeding or stripping of threads. Now your torch is ready for operation. With a Mapp-gas torch, you may have to adjust the fuel regulator.

If you are using a match to light the torch, light the match and twirl or rotate it approximately 180 degrees to create a full flame. Turn the valve control knob to the on position as far as it will go. Hold the

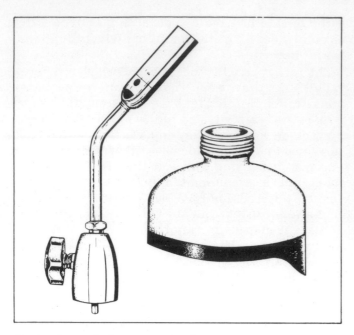

Parts of a typical propane torch.

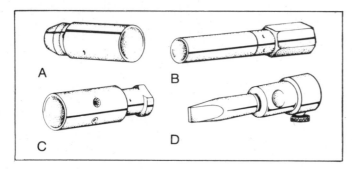

Various types of tips: (A) pinpoint tip, ideal for toys, electrical work, or hobbies; (B) standard pencil-point tip, for a general-purpose flame; (C) brush flame tip, recommended for soft-solder repair and maintenance; and (D) chisel-point soldering tip.

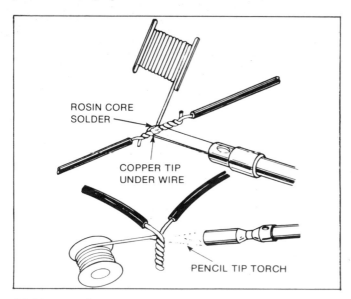

ROSIN CORE SOLDER

COPPER TIP UNDER WIRE

PENCIL TIP TORCH

Making an electric connection with a torch.

match to the burner at the top of and slightly behind the tip. Never hold the match directly in front of the burner tip; the force of the escaping propane or gas will usually blow out the match. When the torch is lit, use the control knob to adjust to the desired flame.

To light the torch with a spark lighter, place the cup of the spark lighter against the end of the burner. Incline the spark lighter about 30 degrees. Turn the valve control knob to the on position as far as it will go. Actuate the spark lighter. Use the control knob to adjust to the desired flame. Always allow the torch to warm up before using it in an inverted or upside-down position.

The operation of the torch when soldering is basically the same as with an iron or gun. Of course, a torch, with its greater heat, can be used for larger soldering jobs, including plumbing work.

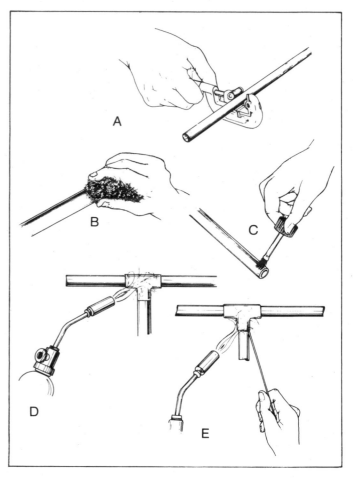

Soldering copper tubing: (A) A tubing cutter gives a clean, square edge when cutting copper tubing; a hacksaw may also be used. (B) Before joining the tubing and fitting, polish the surfaces to be joined with fine steel wool. (C) Apply flux to the end of the tubing, and slip the fitting over the tubing; twist slightly to spread the flux. (D) Apply the heat from the propane torch to the joint, playing the flame so it heats the fitting and tubing evenly. (E) Apply the solder to the joint; the heat in the copper melts the solder which flows smoothly into the joint.

Like many other tools, propane and gas torches are safe when used properly. If abused, they can be dangerous. Keep the following simple precautions for safe torch storage and operation in mind:

1. Do not let unignited gas escape from the torch near any possible source of ignition.

2. Never store propane or Mapp-gas tanks in a confined, unventilated space such as a closet or in any area where the temperature may exceed 120 degrees F. Do not store in a room used for habitation.

3. Never use a flame to test for propane or gas leaks.

4. Never use a tank with a leaking valve or other fitting. If in doubt, test by brushing a generous amount of liquid detergent over the suspected area and looking for bubble formations.

5. Never lay a torch down unless the gas flow has been shut off. If you are maintaining a pilot flame during work pauses, use a rack or stand for the torch, and keep it away from combustible materials.

6. Do not start fires. Be very careful when working near combustible materials, and use asbestos shields when necessary.

7. Never solder a container that holds or has held flammable fluids or gases unless the container has been totally purged of these materials. If in doubt as to the previous contents of a container, thoroughly purge it. Be sure any container you work on is well vented.

8. Propane and Mapp gas consume oxygen and generate toxic fumes; therefore, use a torch only in a well-ventilated area.

9. Avoid breathing the vapors and fumes generated during torch usage. Provide ventilation that will move the vapors away from the work area.

The propane or Mapp-gas torch is a useful tool for auto-body repair, metal sculpture, burning off paint, soldering copper tubing, removing nuts, bolts, and damaged screws, repairing household and farm utensils, thawing frozen pipes, burning weeds, drying spark plugs, and annealing and case-hardening metals. Be sure to follow the manufacturer's instructions for its various uses.

Brazing

As mentioned earlier, brazing is done at high temperatures to join tougher metals. With such metals, ordinary soft soldering will not do, and the soldering iron or gun will not produce enough heat to melt the tougher hard solders. The usual equipment

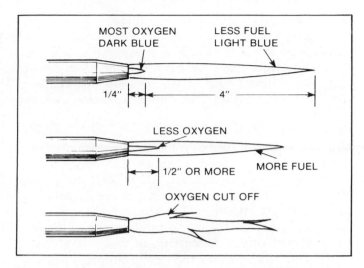

How oxygen—or the lack of it—affects the color of the flame. For most brazing and welding jobs, the most efficient flame is a flame with a sharp light blue outer flame with a pinpoint darker blue inner flame. The blue flame should be about ¼ inch in length. The hottest part of the flame is at the extreme tip of the outside flame. To get a flame of lower temperature, decrease the flow of oxygen and increase the flow of the fuel gas. The inner flame will be longer, ½ inch or more in length, and will also get wider. If you get a long, yellow, pulsating flame, that means that the oxygen supply has been cut off or there is no more oxygen in the cylinder.

used industrially is an oxyacetylene hand torch. The fuel is a mixture of oxygen and acetylene. Both the flame temperature and the amount of heat generated (measured in British thermal units, or BTUs) by the torch depend upon the ratio of oxygen to acetylene, or the fuel ratio. For example, when a 1-to-1 oxygen-to-acetylene mixture is used, a flame temperature of about 5,500 degrees F. is produced. But if we increase the oxygen to a 1.7-to-1 ratio, the flame temperature will increase to approximately 6,000 degrees F. Because the tanks containing the two gases must be kept under pressure, the oxyacetylene torch system requires numerous gauges for proper operation; in various parts of the country, special permits are required by law for the storing of these fuel tanks, so they are not usually found in the average shop. The oxyacetylene system is, however, ideal for mending broken tools, gears, and auto bodies. Joints which have been brazed can stand jars and shocks which soft-soldered joints will not.

In recent years, thanks to the introduction of gas and oxygen weld-braze outfits, the home shop operator can do brazing work. In these outfits, the two necessary gases, oxygen and propane or Mapp gas, are combined in the torch and, when burned at the tip of the torch, produce a flame of intense heat. The temperature may reach as high as 5,300 degrees F. which is high enough to melt most metals, making the brazing and welding of these metals possible. In cutting, a jet of oxygen is supplied in addition to the flame. As a general rule, Mapp gas will produce a somewhat higher heat output and may allow you to complete a braze-weld or fusion-weld job quicker and with less oxygen. The choice of fuels is yours to make.

Oxygen itself does not burn but is an indispensible ingredient in the combustion process. As has already been mentioned, when a highly concentrated source of oxygen is added to the fuel, the flame temperature and rate of heat output are greatly increased. It is impossible, however, to obtain sufficient oxygen from the air to produce the temperatures needed for flame cutting or fusion welding. Therefore the necessary oxygen is supplied from a bottle or in a solid form. That is, the oxygen necessary for the operation of the welder comes from small compressed-oxygen cylinders or is made in a special cylinder from pellets which are used as directed by the manufacturer.

The brazing process is almost identical to welding. The chief difference is that in brazing the two metal pieces are joined by another metal (the brazing rod). In welding, the two metal pieces are fused together. The brazing rod has a lower melting point than the metals to be joined; for this reason, the brazing process is somewhat easier to complete.

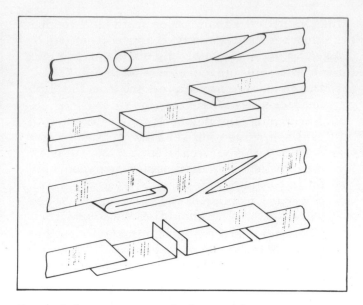

Some of the many ways the lap and butt joints can be made with metal. The fit should be tight; do not depend upon the brazing to fill any gaps.

When the base metal is hot all the way through the joint (not just at the surface), introduce the brazing rod into the torch flame, touching the joint at the hottest point. Rub some flux from the end of the rod onto the joint. When both joint and rod are hot enough, the rod will melt and flow easily and quick-

The Proper Brazing Rods for Various Metals					
Type rod:	Steel	Aluminum, flux-core	Bronze, flux coated	Nickel-silver	Copper-phosphorus
Type torch to use:	Oxygen; Mapp gas	Oxygen; propane or Mapp gas	Oxygen; Mapp gas	Oxygen; Mapp gas	Oxygen; propane or Mapp gas
Apply welding, brazing, or soldering rod:	Melt rod end in molten puddle of base metal	Flux becomes a clear liquid	Rod flows freely on contact with heated metal	Flux becomes a clear liquid, and rod flows freely on contact with heated metal	Rod flows freely on contact with heated metal
Metals*					
Aluminum		x			
Chrome plate			x	x	
Nickel-base alloys			x	x	
Bronze			x	x	x
Copper			x		x
Galvanized iron or steel	x		x	x	
Silver or silver plate					
Stainless steel			x	x	
Steel	x		x	x	
Cast iron	x		x	x	

*Unlike metals can be joined using brazing rods that are suited to both metals. For example, steel and galvanized iron can be joined using steel rods, bronze flux-coated rods, or silver solder.

ly into the joint. You can now move the torch along the joint, repeating the same sequence as required.

As a general rule, bronze brazing is the most versatile and provides the strongest joint. When done properly, bronze brazing is as easy as soldering and produces a bond as strong as welding.

Follow carefully the metal preparation steps. Note that bronze brazing rods often are flux-coated. No further fluxing is required. Heat the metals to be joined until they are cherry red. Try to keep the flame in one place. Moving the flame will allow the metals not in direct contact with the heat source to cool. The tip of the torch should be held about ½ inch from the metal.

When using either the copper-phosphorus or aluminum rods, the procedures are similar. However, the metals will be heated to lower temperatures. Copper-phosphorus rods are used frequently in plumbing work. They allow pipes to be joined even though some water may be in them. When using copper-phosphorus brazing rods, heat the metals until they are a dull red color, then follow the procedures described for bronze brazing.

Aluminum brazing is a bit more difficult since the melting point of the brazing rod is quite close to the melting point of aluminum. For this reason it is best to practice a few times on scraps of aluminum before trying a repair job. Use aluminum flux and clean parts to get a good joint.

When brazing aluminum, heat the joint for about six seconds. Apply the brazing rod. As soon as the rod begins to flow, remove the heat. Repeat this procedure as needed to complete the repair.

Heat control is often necessary to protect the area surrounding the heated metals or to preserve the temperature of the metals to be joined. For example, when brazing pipes that are positioned near wood or other combustible materials, protect these combustibles with a heat shield of asbestos or a similar noncombustible and heat-resistant material. When a large area is to be brazed, the joint should be surrounded with heat blocks or a heat shield to retard heat loss in the metal and reduce the time needed to heat the joint to the desired temperature.

Welding

Fusion welding, which includes gas, arc, and resistance welding, requires that the parent metals be melted. This distinguishes fusion welding from brazing. In fusion welding, as in brazing, the metal pieces to be joined must be thoroughly cleaned of all dirt, paint, oil, and rust. Chemical cleaners may be used, or the metal can be sanded, filed, or cleaned with a grinding wheel. If the metal is not thoroughly cleaned before welding, the joint will be weak. The edges of the pieces to be joined should be trimmed to fit snugly together before welding. Filler or welding rods can be used to fill small gaps, but the strongest weld will be obtained if the metals match well before welding. As the metal is heated with the torch, it will expand, causing the metal to move. It is therefore important to firmly anchor the pieces in position before heating. This can be accomplished with clamps, straps, or bolts. If the joint is not anchored before heating, movement could result in a misaligned repair. Here again practice makes perfect. Since the welding process actually requires melting the metal pieces so they will flow together, it is possible to overheat them and cause distortion near the weld. A few practice welds on scrap metal of the same thickness as the repair job will guide you to the proper application of heat.

Professional welders use one of two welding styles: forehand or backhand welding. In forehand welding, the welding rod is in front of the torch tip, while in backhand welding, the welding rod is in back of the torch tip. Backhand welding tends to result in smaller puddles of molten metal and some welders maintain that this is a more efficient way of welding. But if you find that you can do a better and neater job with forehand welding, use the forehand method. The rules are flexible. In both styles, the welding rod as well as the torch is held at a 45 degree angle to the work.

A great deal of welding can be done without the use of a filler rod. Position the work so that you can start in at the right (if you are right-handed). Hold the torch tip so that the tip points to the left at about a 45 degree angle to the work. Now move the flame across the proposed welding area, back and forth in ¼-inch movements. As the metal starts to melt, keep moving the torch to the left in a sort of zigzag pattern. Do not move the torch faster than the advance of the molten metal. If you move too fast, the weld will not be strong; if you move too slowly, the torch will burn a hole through the metal. As you near the end of the weld area, lift the torch to avoid burning a hole in the metal.

Aluminum melts at a much lower temperature than steel and is rather difficult for the beginner to weld. However, it can be welded—all it takes is a bit of practice. Use scrap aluminum of the same thickness as the work. Successful welding of aluminum is a common, everyday job with the pros—they have the practice, which is all you need. Most trouble with welding aluminum derives from the fact that the operator applies too much heat; about six seconds is usually enough unless working with large castings of aluminum.

It is best to allow the welded joint to cool slowly and naturally. This may take some time, and care should be taken not to touch the metal until you are sure it has cooled. Dipping the metal in cold water will cool the joint faster, but it may also cause distortion of the metal and weaken the joint.

In arc welding, the power is electricity. The torch is actually a holder for a rod of welding metal which is fused to join the melted metal of the pieces being welded. The work is grounded through one wire; the rod completes the circuit; and the very hot arc caused by holding the rod near the work melts the welding rod and welded metal. The arc is so brilliant that an eye shield is required to avoid blindness and to assist the welder in seeing the work as it progresses.

Arc-welding equipment ranges from stationary types that require hundreds of amperes and 240 volts (AC, AC/DC, or DC) to small units that operate on household current. The latter are suitable for light-duty jobs: the joining of rods and plates of ⅛- to ½-inch size. Other arc-welding units are suitable for shop use, especially in the 230- to 300-ampere output range, but such welders require at least a 240-volt, 50-ampere circuit. Also, these larger medium-duty units cannot be considered portable, as are the light-duty units. Of course, dollies make their transportation a relatively easy task.

Most arc welders use a carbon rod, available with the equipment or as an accessory, to braze, repair light-gauge metal, sweat copper tubing, and even construct metal sculptures. Another valuable accessory is an arc stabilizer, which converts any AC or AC/DC arc welder to a versatile tungsten inert gas (TIG) torch. Such a unit will enable you to weld metals such as aluminum, bronze, nickel, and stainless steel, and metals of very light gauge up to ¼ inch thick. The arc stabilizer operates continuously to maintain a long arc. The tip need not touch the metal during welding. This process floods the molten metal with a noncombining inert gas (usually argon) to prevent oxidation and embrittlement of the weld. The TIG torch uses a tungsten electrode; it will not pass impurities on to the weld.

In joining two pieces of metal in standard arc welding, the melted rod forms a bead similar to the melted solder in soldering and in brazing. However, the metals being joined also fuse, and the rod

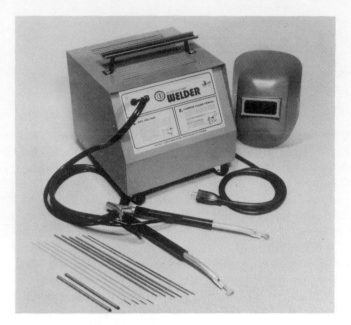

Typical light-duty electric or arc-welding kit.

is fused along with them. Here, too, the rod includes a flux to keep air from the joint and to prevent corrosion and weakness. As in soldering and brazing, the right rod, suitable for the metal pieces to be welded, must be chosen. Obviously, a hard-metal rod will not melt until it attains the right temperature; if the metal pieces to be joined are softer, they will melt before the rod itself.

In comparison with the soldering technique, the welding rod must be moved slowly. If it is moved too rapidly, not enough of the rod will be melted at any one point, and the joint will be improperly filled and weak. Practice is needed, but there is little to learn so you should become skilled after a few practice sessions. Start by joining pieces of scrap in various positions; weld them together, and then try to break them apart to study your work. If you succeed in breaking them apart, your failure will be visible. If not, you have learned all you need to know with the exception of appearance. Neat welding joints, while not important from the standpoint of strength, mark your work as that of an expert.

Because of design variations in brazing and welding equipment, it is not possible to give full instructions on how to use these tools. Fortunately, the manufacturers do supply good instruction manuals which should be carefully followed.

Fastening with Staples and Rivets

Two of the more popular methods of fastening are stapling and riveting. The former provides an efficient method of attaching insulation, roofing material, building paper, screening, underlayment, weatherstripping, upholstery material, ceiling tile, and many other products to wood and composition surfaces.

Riveting is used extensively for joining and fastening metal sheets when brazing, welding, or other locking techniques will not provide a satisfactory joint. But rivets can also be used for joining fabrics, plastics, leather, and sometimes wood to metal. They are easy to install and make an almost instantaneous, tight, permanent joint.

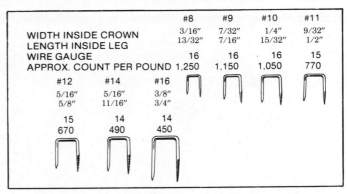

	#8	#9	#10	#11
WIDTH INSIDE CROWN	3/16"	7/32"	1/4"	9/32"
LENGTH INSIDE LEG	13/32"	7/16"	15/32"	1/2"
WIRE GAUGE	16	16	16	15
APPROX. COUNT PER POUND	1,250	1,150	1,050	770
	#12	#14	#16	
	5/16"	5/16"	3/8"	
	5/8"	11/16"	3/4"	
	15	14	14	
	670	490	450	

Flat wire or U-shaped staples come in lengths from $^{13}\!/_{32}$ to ¾ inches.

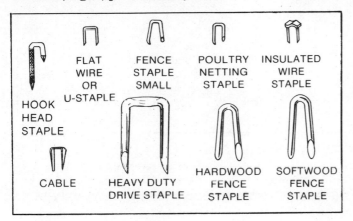

Typical hand-driven staples.

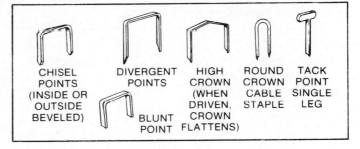

Typical power-driven staples.

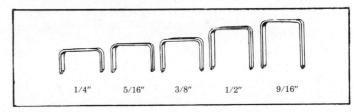

1/4"   5/16"   3/8"   1/2"   9/16"

Actual staple sizes.

Stapling

Staples are two-pointed fasteners made of wire. They can be driven by hand or by mechanical staplers. When fastening with the hand-driven types, a hammer is used to drive them in much the same manner as nails and tacks.

Today, most staples are driven by mechanical staplers. While these staples are available in several types, most manufacturers produce them for average use in five standard leg lengths from ¼ to $^9\!/_{16}$ inch. Which length you use depends principally on the thickness of the materials being stapled, but remember that too short a staple for the job may result in the staple pulling out. On the other hand, too long a staple may go through the material, split it, or meet resistance.

The number of staples used to hold a given project depends upon the weight of the material and the stress placed upon the staples in use. For exterior or special-purpose use, many hardware stores and home centers have special or corrosion-resistant staples for your gun. Keep in mind that most staplers will take staples produced only by the manufacturer of the tool. Staples made by other manufacturers will not fit. Thus, it is important to select a good stapler and one that will take all the sizes and kinds of staples you wish to use.

Mechanical Staplers There are three basic types of stapling tools that have applications in the average shop or around the home. They are: (1) the staple gun; (2) the hammer tacker; and (3) the plier

Staple Uses by Leg Length				
¼-inch leg	5/16-inch leg	⅜-inch leg	½-inch leg	9/16-inch leg
Light upholstering, shelf trimming, screens, window shades, decorations, valances	Thin insulation, storm windows, draperies, upholstery, heavy fabrics	Weatherstripping, roofing papers, light insulation, electrical wires, wire mesh	Canvas, felt stripping, underlayments, carpets, fiberglass	Ceiling tile, fencing, insulation board, metal lathing, roofing

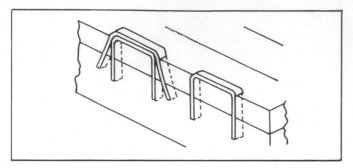

When fastening ceiling tile, a double 9/16″ chisel-point staple is sometimes employed for better holding. A special attachment to align and steady the second staple is available for some guns.

stapler. The latter is used primarily for fastening thin sheets of material together. Both sides of the material must be accessible, since the tool bends the staple legs closed on the reverse side.

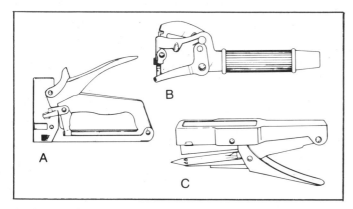

Basic stapling tools: (A) staple gun, (B) hammer tacker, and (C) plier stapler.

Of the two tools that are exclusively used for stapling, the staple gun is the most commonly used. While specific light-, medium-, and heavy-duty staple guns are available, many of the newer ones have variable power drives. That is, by turning a knob or moving a lever, you can change the stapler from light duty for soft surfaces to heavy drive for hard ones, or vice versa. In recent years, the electric staple gun has entered the picture. This push-button operated tool makes it very easy to drive the longest of staples into a hard surface.

There are specialty staple guns that are used to fasten electric wires, cables, and copper tubing up to ½ inch in diameter. Still another staple gun shoots a flared staple which enters soft material and locks inside without penetrating through the material.

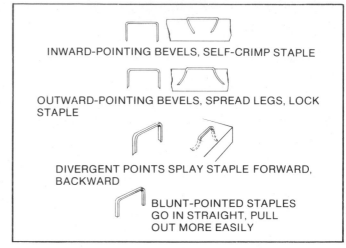

INWARD-POINTING BEVELS, SELF-CRIMP STAPLE

OUTWARD-POINTING BEVELS, SPREAD LEGS, LOCK STAPLE

DIVERGENT POINTS SPLAY STAPLE FORWARD, BACKWARD

BLUNT-POINTED STAPLES GO IN STRAIGHT, PULL OUT MORE EASILY

How different staples hold.

Some staple gun kits feature attachments to make them more versatile. For instance, there may be a window shade attachment to fit round shapes, another to keep screening taut while stapling, and a centering device for making small wire fastening easy. Most kits also contain a staple lifter for removing staples; in some guns this is a built-in feature.

Although hand-powered staplers are not hard to use, the new electric models are effortless. You simply touch the trigger at the front of the grip and, about as rapidly as you wish, it drives home a staple.

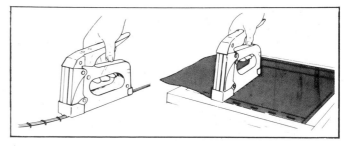

Your stapler can be made more versatile by using an electric line attachment (left), and a special screening attachment (right).

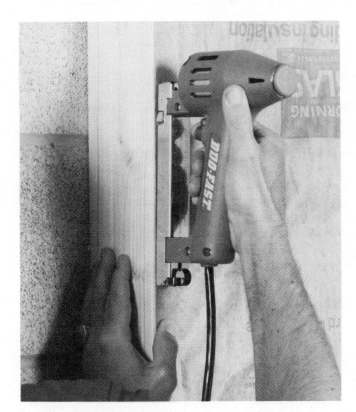

Electric staplers make stapling effortless.

There is a conveniently located slide-button safety to lock the trigger button so it cannot be discharged accidentally.

To load a typical staple gun, proceed as follows:

1. With the operating lever or handle locked, hold the gun in the normal firing position. Push the loading latches forward on both sides simultaneously, and pull out on the staple channel with the other hand.

2. Holding the staple gun upside down, pull back on the staple channel until the spring pulls the feeding bar completely back. Then put the staples in the staple pocket with the legs pointing up.

3. Lower the staple channel, pull back on the loading latches, push the channel securely into place, and release the latches, making sure the channel is secure.

4. When you are ready to use the gun, release the handle lock and allow the handle to open. Your staple gun is now ready to fire when you depress the handle. When firing staplers, you will get more power by pressing down on the front end. When driving is extremely tough, pressing down on the nose with the palm of the hand will reduce recoil.

The above procedure may vary slightly with different makes of staple guns. For example, many staplers are equipped with handle locks or safety devices. In any case, remember that a staple gun is not a toy, and it can fire staples into the air. Always keep a stapler out of the reach of children. Before you start to squeeze the handle, be sure the gun is firmly in position.

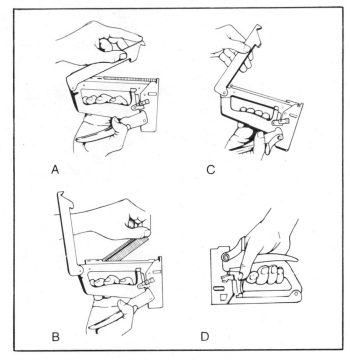

The four steps of staple gun operation: (A) open, (B) load, (C) close, and (D) fire.

To use the hammer tacker, swing it like a hammer; when it hits the surface, it automatically drives a staple. While you do not have accurate aim with a hammer tacker as you do with a staple gun, the jobs it is used for—tacking up building paper, vapor barriers, and insulation, as well as attaching shingles, underlayment, and so on—do not require it. The hammer tacker is generally loaded in the same manner as the staple gun.

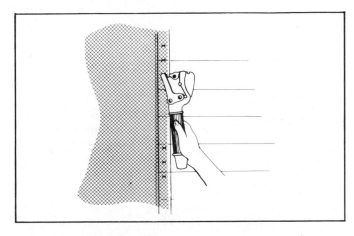

A hammer tacker in use.

Brad Nailers   Brad nailers or nail guns look and basically operate in the same manner as staple guns, except that instead of staples they drive small brad-type nails, usually about $1\frac{1}{32}$ inches long (18 gauge). The nail gun operates on the lever principle, lifting an internal ram that is driven forward with great force by a ram spring. (The safety guard locks the ram.) The nail driver is only slightly larger than the nailhead, making countersinking nearly invisible. Each nail is placed precisely and accurately

An electric brad nailer in use.

with a single blow of the hardened steel driver. This means you can nail at much faster speeds with no surface damage, no mars, and no hammer dents. These devices are especially useful for paneling, fine finishing, curved molding, and hard-to-get-at places in corners or near floors and ceilings. Some brad nailers are electrically driven. The nails are available in wood-tone colors and in off-white.

Riveting

The major types of extensively used rivets include the standard type and pop rivets. Standard rivets must be driven using a bucking bar, whereas pop rivets have a self-heading capability and may be installed where it is impossible to use a bucking bar.

Standard Rivets   The basic standard rivet could be considered a threadless bolt which is slipped through a hole drilled in the materials that are being put together. When struck with a rivet hammer, the shank is flattened into a mushroomlike head that keeps the materials from coming apart. For thin or lightweight materials, a washer may be slipped over the shank for extra holding power.

Wherever possible, standard rivets should be made of the same substance as the materials being joined to reduce corrosion. For example, use aluminum rivets for joining aluminum; copper for cop-

per or brass; steel for steel. For joining nonmetals, employ any rivets you like, always keeping in mind the problem of corrosion action between unlike metals.

Rivets are also classified by lengths, diameters, and their head shape and size. Selection of the proper length of rivet is important. If too long a rivet is used, the formed head will be too large, or the rivet may bend or be forced between the sheets being riveted. If too short a rivet is used, the formed head will be too small or the riveted material will be damaged. The length of the rivet should equal the sum of the thicknesses of the metal plus $1\frac{1}{2}$ times the diameter of the rivet.

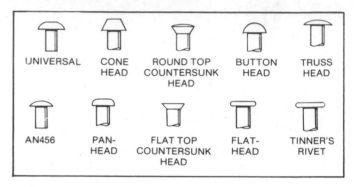

Some common types of rivets.

Before setting any standard rivets, clamp the pieces of material to be joined together and center-punch the hole location. Then, drill a hole just slightly larger than the rivet. After the holes are drilled, remove any burrs.

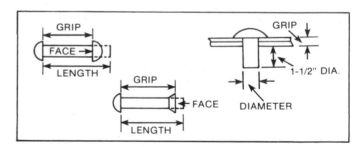

What is meant by the "grip" of a rivet.

Guide for Selecting Rivet Sizes for Sheet Metal Work	
Gauge of Sheet Metal	Rivet Size (weight in pounds per 1000 rivets)
26	1
24	2
22	2½
20	3
18	3½
16	4

The riveting procedure itself involves three operations: drawing, upsetting, and heading. Insert the rivet in the hole in the materials and place the rivet head down on an anvil. The sheets are drawn together by placing the deep hole of the rivet set over the protruding rivet shank. Strike the head of the set with a rivet hammer.

Upon removal of the set, the head of the rivet is struck lightly to upset the end of the rivet. Finally, the heading die (dished or cupped part) of the rivet set forms the head of the rivet when the hammer again strikes the head of the rivet set to complete the

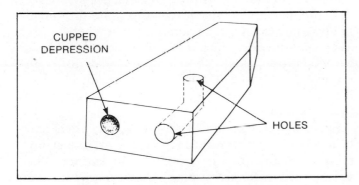

Drawing, upsetting, and heading a rivet.

head formation. If you do not want the original rivet head flattened by the anvil, place it in the dished portion of a second rivet set. Of course, if you are using flathead rivets, the heading step of the riveting operation can be omitted.

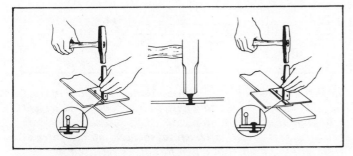

A typical rivet set.

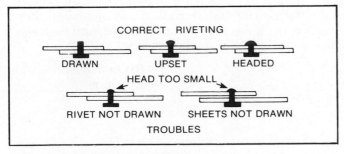

Correct and incorrect riveting.

It is also possible to install rivets without using a rivet set. The operation takes a little longer and the

end results may not be as professional looking as when using the set. With the rivet in place on the anvil and the materials you are joining tightly together, blunt the end of the protruding shank with a sharp hammer blow. Then, carefully peen it over, making a studied effort to make the end flare out evenly to all sides and to bring it to a half-moon contour. It is sometimes necessary to tap the newly formed head with the flat end of the hammer to tighten the joint.

Pop or Blind Rivets Pop or squeeze-type rivets have a major advantage over standard rivets: they can be used for blind fastening. This means that they can be used where there is limited access or no access to the reverse side of the work.

The rivet is made up of two parts, the mandrel and the rivet body. To determine the proper size of the pop rivet, first determine the size of the predrilled hole. Use the rivet that matches the size of the hole; for example, use a ⅛-inch rivet for a ⅛-inch hole. On jobs requiring the drilling of new holes, the rivet diameter is best determined by the strength requirement. Pop rivets are usually available in steel and aluminum, and in a variety of diameters (usually from ⅛ to ¼ inch) and lengths that can be used for

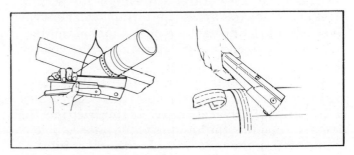

Typical uses of the pop riveter.

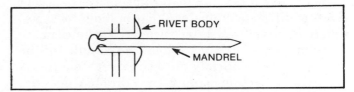

Parts of a typical pop or blind rivet.

work up to ½ inch thick. Steel rivets are much stronger than aluminum ones. Use steel rivets for very-heavy-duty jobs and when riveting steel to steel. Employ aluminum rivets for lightweight jobs such as aluminum to aluminum, fabrics, plastics, and so on. A few companies make copper rivets which should be used for repairing and fastening copper items. Neither copper nor aluminum rivets will rust.

There are two basic designs of pop rivets, closed end and open end. The closed-end rivet fills the need for blind rivets which seal as they are set. They are

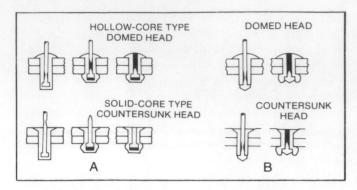

Popular blind rivets: (A) closed-end type and (B) open-end type.

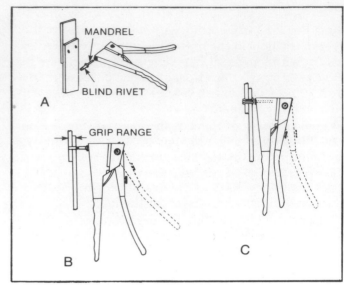

The three steps in blind riveting: (A) inset mandrel in rivet tool so head of rivet is flush with face of tool, (B) slip rivet through hole drilled in materials and squeeze tool handle until rivet is set, and (C) after a final squeeze, mandrel breaks off. It will fall from tool when handle is released.

gastight and liquidtight when used properly, since a high degree of radial expansion provides excellent hole-filling characteristics, and the mandrel head is within the core of the rivet body.

The open end is not liquidtight because the mandrel head which remains in the rivet body is not enclosed within that body as in the closed-end type. This obviously leaves room for possible seepage of liquid.

There are special rivets with extra-large flanges for large holes in soft materials, countersunk rivets for flush surfaces, threaded rivets for setting threaded holes in various materials, washer-type rivets that provide back-up plates for soft materials, and white rivets for repairing white-colored objects.

The operation of the riveter (often called a riveting plier) is simple. After the holes have been drilled in the parts to be riveted, the rivet is inserted into the riveter by opening the handles of the plier completely and setting the pointed end of the rivet into the hole of the nosepiece as far as it will go. (Most pop riveters have interchangeable nosepieces to accept rivets of different diameters.) The bulbous head portion of the rivet is inserted through the holes in the material being fastened. Then, the handles of the tool are squeezed firmly until the rivet stem is broken off. This action is repeated if the stem does not separate on the first squeeze. The squeezing action helps to flare out the rivet on the other end.

To remove the separated stem from the tool, open the handles completely and turn the tool over. The separated stem will fall out of the riveting pliers. (Caution: Be sure to remove the separated stem so the riveter will accept the next rivet and to prevent the separated stem from being expelled into the air.)

When rivet holes become enlarged, deformed, or otherwise damaged, use the next larger size rivet as a replacement or use a backing plate for added holding power. Metal backup plates or washers are also recommended for use in soft materials such as canvas, fabric, plastic, and leather. Incidentally, to remove a pop rivet just drill out the head. There is little danger of damaging the work, as in removing most standard rivets.

Several attachments for standard vise pliers can be used to clinch standard rivets, snaps, and grommets in metal, plastic, canvas, and leather. The various anvils for rivet, snap, and grommet fasteners are attached to the jaws of the vise pliers. For snaps and grommets, the tool punches the material and clinches the fastener all in one operation. For riveting, the holes must first be drilled, and then the rivet is clinched by the tool.

Attaching It on the Wall

Sooner or later every homeowner is faced with the problem of how to attach something to a wall—a shelf, a drapery rod, a cabinet, or even a picture. Of course, the easiest way to put anything on the wall is to use the correct fastener for the job.

There are several basic considerations with which the homeowner must deal when choosing the proper type of fastener to attach an item to a wall. The first is the wall itself. For jobs around the home, the homeowner will be dealing with hollow walls—dry wall or plaster construction—and masonry walls. Fasteners for hollow walls actually go through the outside surface and anchor themselves on the inside of the back surface. On the other hand, solid and masonry wall fasteners actually anchor themselves against the inside of the hole into which they are placed. The shell expands and grips the inside as the fastener is tightened. Some of the fasteners usually used for masonry walls can also be used for hollow walls.

In addition to the wall material itself, the homeowner must consider the general conditions under which the fastener will be installed and used. Will it be necessary to remove the fastener from time to time? How will the anchors be spaced? How deep will the hole in the wall be? All of these questions will aid in determining the type of fastener to be used.

Not only is the type of fastener important, but the size and number must also be taken into account. When determining the size and number, the most important consideration is the weight of the object to be held. This along with three other factors constitute the maximum working force of the fastener: the type of loading, the angle of loading, and the way in which the load is placed on the fastener.

There are generally two forces which affect the holding power of the fastener to be used: shear and combination force. Shear is the downward pull exerted on the fastener by the item it is holding. For example, a picture hanging from a nail which has been horizontally driven into the wall would constitute shear force. The only consideration you would have would be whether or not the weight of the picture would bend the nail. This, of course, is true if you assume that the material into which you are driving the nail will not crumble.

Combination force, on the other hand, consists of an outward and a downward pull by the load. A bookshelf supported by a bracket exemplifies this type of force. It is necessary for the bracket in this case to not only hold the load up, but to keep it from pulling out as well. A load of this kind obviously requires a different fastener (kind and size) than the picture hanging on the wall.

In addition to the shear and combination forces described, there is another very important factor which will determine the decision you will reach in choosing a fastener—vibration. Will the load being held remain static, or will it be vibrating at any time? The best example of a vibrating load is an air conditioner. When the air conditioner is turned on, it vibrates. This vibration would cause a fastener such as a nail to eventually loosen up and may even cause it to pull out. In a case such as this, it would be best to use some sort of screw-type fastener.

One final element which may affect your choice of fastener is the thickness of the wall. To determine the thickness of a hollow wall, insert a piece of stiff wire which has been bent at one end through a hole drilled in an inconspicuous area. When the hook in the wire catches on the back face of the wall, mark the wire at the hole opening. Then, withdraw the wire and measure the distance. This is the thickness of the wall and allows you to determine what size the fastener will have to be to properly anchor.

When taking all of these factors into consideration, you will be able to determine the type, size, and number of fasteners to use for the job. The fastener must be able to support the heaviest load you will ever place on it and do it with a reasonable margin of safety. The basic rule to keep in mind for providing a safety factor is: never use a safety factor of less than 4 to 1 for general fastening, and whenever extreme vibratory conditions are involved, use a 10 to 1 ratio. In other words, if you want to hang something on the wall which weighs 25 pounds, use a fastener with a maximum holding power of 100 pounds. This will provide a safe and secure mount.

All-Purpose Wall Fasteners

As already mentioned, there are several all-purpose fasteners that can be used in both hollow and solid walls and in almost all materials from wood to concrete. There are three basic types: plastic anchors, nylon expansion anchors, and nylon drive anchors.

Plastic Anchors The most economical of wall fasteners, plastic anchors are used for mounting any fixture normally held by wood screws—large pictures, mirrors, kitchen and bathroom accessories, drapery hardware, and so on. The anchor itself consists of a hollow sleeve which is inserted in a hole drilled in the wall. The hole in the center of this plug is tapered so that it is narrower at the tip than it is at the surface. A lip prevents the anchor from slipping through and covers the raw edges of the hole. The hole should be the same diameter as the shank of the fastener. Then, insert the wood screw through the fixture and drive into the anchor. The screw expands the back end open to wedge it tightly inside its hole. Plastic anchors are available in various lengths and diameters to accept various sized wood screws and to handle varying loads.

When using plastic anchors, it is important to remember that hole size is critical—the anchor must fit snugly and is only as good as the wall material. It will not hold against much horizontal pull, especially in crumbly masonry.

Plastic Anchors—Size and Load Capacity		
Screw Size	**Anchor No. & Length**	**Load (pounds)**
4, 6, 8	#1 (⅞″)	650
10, 12	#2 (1″)	850
14, 16	#3 (1½″)	1075

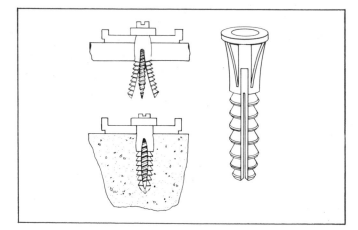

The installation of a plastic anchor.

Nylon Expansion Anchors This type of anchor expands as the screw is tightened, and sets itself firmly into both hollow and solid walls. It is available generally in one size: 1 inch long which requires a ⅜-inch hole. These anchors usually come

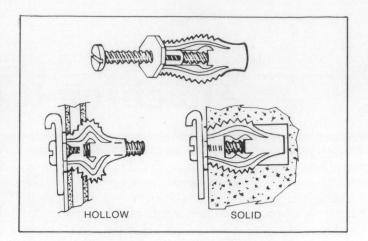

The installation of a nylon expansion anchor.

with a 1⅛-inch self-tapping screw and are able to hold shear loads of 280 pounds in concrete and 50 pounds in ⅜-inch thick wallboard.

To install, drill a hole the same size as the sleeve and at least 1½ inches deep. Then, pinch the anchor between your fingers and tap it in place with a hammer. Insert the screw through the object to be held into the anchor and tighten. These anchors are not satisfactory in hollow walls more than ⅜ inch thick, because they do not expand behind most plaster walls.

Nylon Drive Anchors The shank of the nylon drive anchor bulges as the screw is tightened to set itself firmly in both hollow and solid walls. (The standard drive anchor cannot be more than ⁹⁄₁₆ inch thick.) The so-called standard type is available in one size. It has a ¼-inch shank diameter, ¹⁵⁄₁₆ inch long, and has a load capacity of 370 pounds in concrete, 440 pounds in brick, and 470 pounds in cinder block.

To install the drive anchor, drill a hole the same diameter as the sleeve. In solid walls, drill slightly deeper than the length of the screw. Push the anchor's sleeve in up to the flange and hold the fixture against it while tightening the screw through the mounting hole.

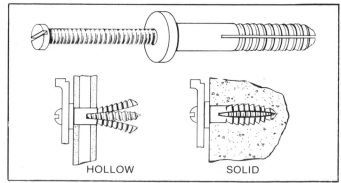

The installation of a standard-type nylon drive anchor.

A recent innovation in nylon drive anchors is the nail-in anchor. This anchor is about ¼ inch in diameter but is available in various lengths of less than 1 inch up to over 2 inches and in three head styles: round, mushroom, and flat. To install a nail-in anchor, drill a hole the same size as the anchor. Set the anchor, with the fixture attached, in the hole, insert the nail, and drive it flush with the top of the anchor. If it is necessary to move the fixture, unscrew the threaded nail, and remove the anchor body with great ease.

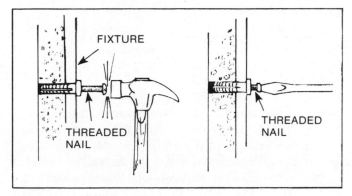

Steps in installing a nail-in drive anchor.

Adhesive Anchors If you do not want to make holes in a wall, adhesive anchors can be used. They are flat perforated steel plates that have either a nail or a bolt projecting from the face. Simply spread the adhesive on the anchor and press it to any solid wall. After the adhesive has set, you can attach a fixture or lumber to the anchor by hammering or bolting it on.

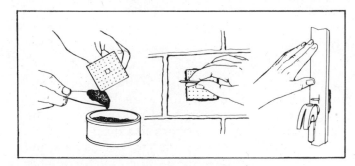

How an adhesive anchor is installed.

Fasteners for Hollow Walls Only

In hollow wall construction, the holding power of a fastener depends primarily on the strength of the wall material; dry wall is weaker than plaster. It should be remembered that in hollow wall construction, a few hundred pounds of weight is usually the maximum.

If the studs are located, they are fine mounting spots for both nails, screws, and large bolts. But, if any of these fasteners should break in between the studs, they will leave some cracked plaster or a useless hole in the wallboard. To locate the studs, use any of the following methods:

1. Tap the wall lightly at the point you wish to drive the fastener. A solid sound indicates the presence of a stud, while a hollow or echoing sound means that no stud is present.
2. Another method of checking is to drill a few small holes (about ⅛ inch) in an inconspicuous place. Continue to drill until you bore into a stud.
3. Take off the baseboard and check to determine where the two dry wall panels meet. This point will indicate the center of a stud.
4. There are several stud finder devices on the market which, when used as directed by the manufacturer, make the job of locating studs fairly easy.
5. Studs are generally located either 16 or 24 inches on center, depending on local building codes. Try measuring out from the corner of the room to find the approximate location of a stud, then use any of the above techniques to find the exact location of the stud.

When hanging heavy mirrors, pictures, storage cabinets, and bookcases, it is a good idea to support them by means of a crosspiece bridging the studs. Use at least Number 14 screws or lag bolts to secure the cross member to the studs.

Where the location of the studs does not lend itself to fastening objects, the following fasteners may be used for hollow wall construction.

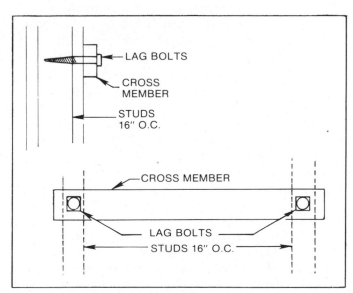

Fasten heavy objects to a bridge piece between the wall studs.

Toggle Bolts   Toggle bolts are particularly good on soft, crumbly material, such as acoustical plaster. They are usually available in three head styles: round-, flat-, and mushroom head, with zinc plating to prevent discoloration of the material in which they are used. They can be used to provide positive anchoring to hollow tile, building block, plaster over lath and gypsum board, and various types of dry wall construction.

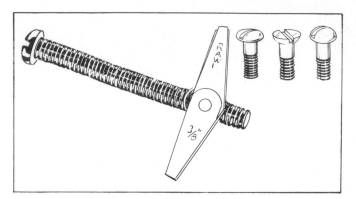

The spring-wing type toggle bolt.

Toggle Bolts—Size and Load Capacity

Length (inches)	Bolt Diameter (inches)	Load (pounds)
2, 3, 4	1/8	200
2, 3, 4, 5, 6	3/16	450
3, 4, 5, 6	1/4	925
3, 4, 5, 6	5/16	1150
3, 4, 5, 6	3/8	1500
4, 6, 8	1/2	1800

To install a standard toggle bolt, drill an oversized hole to admit the wings when folded. Then, proceed as follows:

1. Insert the bolt through the item to be fastened, turn the wings, then fold them back completely. Insert the bolt through the hole until the wings spring open inside the wall.
2. Pull back on the assembly to hold the wings against the inside wall; this prevents the wings from spinning while tightening the bolt. Turn the bolt with a screwdriver until firmly tightened.
3. When the installation is complete, the wings of the standard toggle bolt form a perfect 90 degree angle with the inside wall surface. The entire length of the wings is supported by the material, which provides a full wall bearing distribution of the load.

There are several special types of toggles. For instance, the toggle fixture hanger is a time and

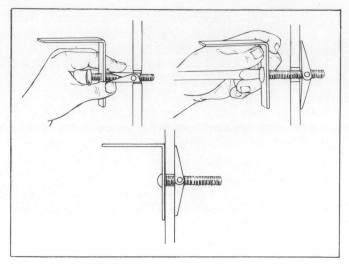

Installation of a toggle bolt.

money saver during extensive remodeling work. The toggle fixture hanger spreads the load over 10 to 12 inches and is equally suited for fastening junction boxes and hanging lighting fixtures. The 3/8-inch nipple fits through the knock-out hole, and a lock nut then tightens the fixture in place. To remove this type of toggle bolt, insert a screwdriver in the nipple and push the key up. Then, pull it down to collapse the wings, and remove it.

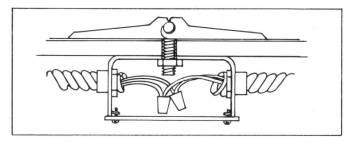

A giant toggle fixture bolt.

The tie-wire toggle bolt is another fastener specially designed for remodeling work. Primarily used for suspension of acoustical ceilings (where heating and air conditioning ductwork has been installed), the tie-wire toggle comes complete with a spade end bolt, standard toggle wing, 1½-inch fender washer and hexagon nut.

There are also several variations of the toggle principle. The gravity toggle, for example, works on the wing principle, but is installed in a slightly different manner. They are available in three bolt sizes —1/8, 3/16, and 1/4 inch—and in lengths for 3/8-inch gypsum board to masonry block. Their recommended working loads vary from about 25 to over 350 pounds, depending on the thickness of the material through which they are fastened. They are installed as follows:

1. Push the gravity toggle through a drill hole (⅜-inch hole for ⅛-inch bolt, ½-inch hole for 3/16-inch bolt, and ⅝-inch hole for ¼-inch bolt).
2. Grasp both strap ends firmly and pull the toggle snugly against the inside wall.
3. Slide the washer evenly along the straps until the inner rim of the washer seats completely inside the drilled hole.
4. Snap off the plastic straps flush with the washer.
5. Insert the screw through the fixture, engage the toggle bolt thread, finger tighten, and secure with a screwdriver. The screw can always easily be removed.

Except for a few special types, such as the fixture bolt, the major disadvantage of the toggle is that if a fixture is removed, the wings drop off and are lost in the wall. Also, installation is awkward where a fixture requires several bolts, since all must be inserted through the fixture before positioning.

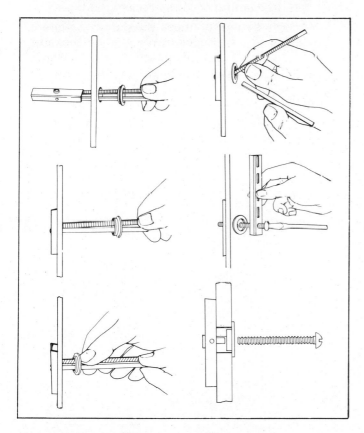

Steps in the installation of a gravity toggle.

Wall Expansion Anchors—Molly Bolts   This anchor expands inside a hollow wall and grips the back of the wall. Unlike the standard toggle bolt, it has a sleeve which remains in the wall and permits withdrawing and replacing the bolt. It also has a lip which covers the edges of the hole. Unless this fastener is correctly sized for the wall thickness, it will

not work. Insert a crochet hook or bent wire into the wall to determine the thickness of the wall.

	Expansion Anchors—Size and Wall Thickness	
Size	Overall Length (inches)	For Wall Thickness (inches)
XS	¾—⅞	0—¼
MS	1—1½	1/16—½
S	1½—2¼	⅛—¾
L	2—2¾	⅝—1¼
XL	2½—3½	1¼—1¾

Wall expansion anchors will accommodate weights of 50 to 500 pounds, depending on the overall bolt length. To install this type of anchor, use the following procedure:

1. Drill a hole using the drill size specified on the anchor's package. Insert the fastener. Tap it with a hammer until the prongs are embedded in the wall and the head is flush with the wall.
2. With a screwdriver, turn the screw until a definite resistance is felt. This indicates the expansion anchor is set.
3. Remove the screw. Place the object to be fastened in position and tighten the screw.

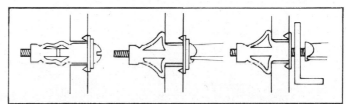

Steps in the installation of a Molly-type expansion anchor.

Some expansion anchors are available that require no drilled hole. Simply drive this type of anchor through the dry wall or plaster and remove the screw to set the unit.

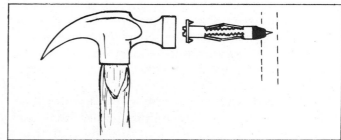

A nondrill hollow wall anchor.

While most expansion anchors are made of metal, a few are made of plastic. To install this type, proceed as follows:

1. Make a hole in the wall of the size recommended for the fastener. Then, fold the fastener and insert it into the hole.
2. Using a hammer, tap the anchor flush to the wall. Engage the fastener with the tool provided by the manufacturer and push till the anchor pops.
3. Place the object to be fastened in position and tighten the screw.

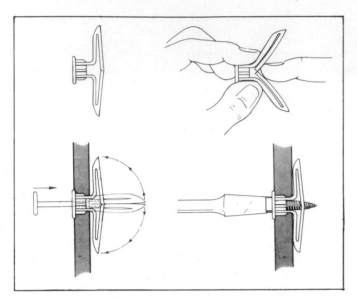

Installation of a plastic expansion anchor.

Picture Hangers   There are a great many picture hangers on the market that can be used on hollow wall surfaces. For light-duty tasks, such as hanging small pictures and decorative plates, there are hangers available which can be cemented in place against the wall. Some of these are adhesive-backed, while others come with a separate liquid adhesive. Since this type of fastener adheres to the surface only, its load capacity is limited by the strength of the surface coating (wallpaper or paint) already on the wall.

When greater strength is required to support larger pictures and mirrors, angled picture hangers are often used. These hooks are designed so that they hold a nail at an angle which will provide the greatest holding strength for its size in dry wall or plaster walls. They are available in various sizes and will support objects up to 100 pounds.

For this type of fastener, place the hanger flat against the wall. Insert the nail through both holes of the hanger and drive the nail into the wall. It is wise before driving a nail into plaster to make an X with two pieces of masking tape at the point where the nail is to enter the wall. This helps prevent plaster from chipping.

To hold the picture itself, either picture wire and

screw eyes or picture self-leveling hangers can be used. These items can be purchased at your local hardware store or home center. With either fastener, remember that a single picture should be hung so that the center of the picture is at eye level with an average person standing in the center of the room. It is always better to locate it a little lower than higher.

Fasteners for Use in Masonry

With the expanded use of masonry in construction, the importance of masonry fasteners is increasing. In all homes, a garage, basement, or patio requires some kind of masonry anchor. It should be kept in mind that the holding power of a fastener in a solid masonry or concrete wall depends to a great extent on the strength and tightness of the material. In solid walls, fasteners are able to withstand up to several hundred pounds of pull and weight capacity.

The successful use of most masonry fasteners depends on two factors: the proper hole and correct anchor. The easiest way to achieve a proper hole in brick, stone, and concrete is to use a carbide-tipped bit and a portable electric drill (at least a ⅜-inch size). When you use these hard-tipped bits, however, take care not to break or dull the tip. Carbide tips are brittle and breakable; do not ram the drill hard

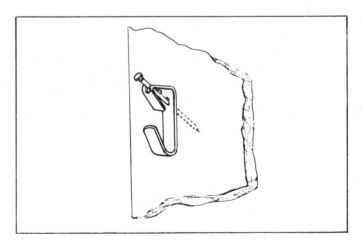

Installation of a picture hanger.

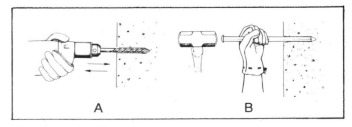

*(A) Using an electric drill and a carbide-tipped bit.
(B) Hand drilling with a star drill.*

against the work. Maintain a steadily increasing pressure as the bit enters the work. Do not let the drill ease up or run idly in the hole. Once the hole is started, increase the pressure as the bit cuts away at the material. If you should get tired (or to free the bit of debris), pull it out of the hole for a minute, but do not let up on the pressure while the drill is working. Actually, for all practical purposes, the amount of pressure that should be used depends directly upon the hardness of the material being drilled. Coolants are not necessary, as a general rule. It may be desirable, however, when hard material such as tile or porcelain is being drilled, to use water or turpentine as a coolant. When a coolant is used, the entire drill point should be kept wet.

Keeping the hole free of dust is important when using a masonry drill. A drill having a flute or twist along the body will itself remove the cutting dust; however, when excessive moisture is encountered, particularly in horizontal or downward drilling, it may be necessary to lift the drill slightly from time to time to clean the hole. If the hole is deep, blow it out from time to time (a syringe or pump is best to avoid getting dust in the eyes), or flush it out with water from a small syringe.

The drilling of small holes is never a problem, whether the material is hard or soft. Large-diameter holes, however, should first be drilled to a small diameter and then either to a larger or to the desired diameter, depending upon the hardness of the material and the diameter of the desired hole. When drilling holes in reinforced concrete, examine the hole occasionally to avoid drilling through large pebbles. If a pebble is encountered, it is best to break it with a center punch and hammer to protect the drill point.

Hand drilling is also possible, but it takes a great deal longer. To accomplish the task, you will need a star drill and a heavy drilling hammer or sledge. The cutting end of the star drill resembles four chisels joined at their edges to form a cross (star).

A star drill should be struck squarely with the hammer or sledge, and it should be rotated after each blow as it goes through the masonry. Every so often blow out the dust. Continue until the desired depth is reached. A piece of tape wrapped around the drill will help you to know when you have made a hole of the proper depth. Always wear goggles and gloves when drilling in concrete.

There are many different types of fasteners that can be used with masonry materials. Standard toggle bolts can be used in concrete-block walls, but only if the object you are hanging is placed over one of the hollow cores in a block. Once the bolts are secure, you cannot ask for a stronger anchor, because nothing can pull them out unless you unscrew the bolts from the spring-wing nuts hidden within the blocks. Expansion anchors that are designed for hollow wall use can also be used with hollow core blocks.

Of course, wood dowels can be used as masonry fasteners. First, make a hole with either a carbide-tipped bit or with a star drill. Using a slightly oversized, greased dowel, drive it in flush with a hammer. Then, drill a pilot in the center of the dowel slightly smaller than the holding screw. To complete the installation, place the object to be fastened in position and drive the screw home.

Most masonry fasteners, however, are designed specifically for masonry work. The most popular types are described here.

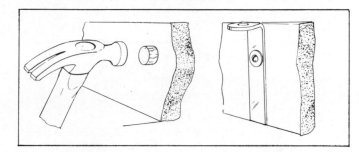

Using a wood dowel as a masonry fastener.

Fiber Plugs A fiber plug is a masonry anchor designed for use with wood, sheet metal, and lag screws. It is formed of braided jute, nature's toughest fiber, compressed into a tubular shape, and treated with chemicals and heat to give it a tough and elastic binding. A lead lining makes it possible for the screw to reproduce its own thread and keeps

Fiber Plugs—Size and Load Capacity		
Screw Size	Anchor Length (inches)	Load (pounds)
5-6	⅝, ¾, 1	550
7-8	⅝, ¾, 1	885
9-10	¾, 1, 1¼, 1½, 2	1150
11-12	¾, 1, 1¼, 1½, 2	1526
14	1, 1¼, 1½, 2	1590
16	1, 1½, 2	2150
20	1, 1½, 2	2830
22	2	3500

the jute fibers from being cut by the screw. Fiber plugs are made in a complete range of sizes to match most screw sizes.

Installation of the fiber plug is very simple; drill the proper size hole through the fixture, insert the fiber plug, and turn the screw home. Since the plug depends on tremendous compression for its holding power—the flexibility of the jute permits it to be forced and compressed into every necessary irregularity in the walls of the hole—it is absolutely necessary that the plug, the drill, and the screw sizes are all the same. For instance, using a Number 12 screw, drill the hole with a ¼-inch drill and use a Number 12 fiber plug.

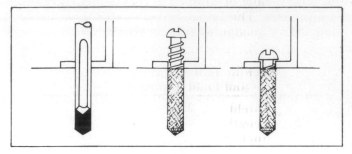

Installing a fiber plug.

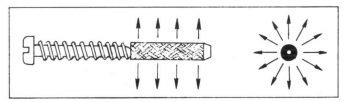

How a fiber plug holds.

The shoulders of the wood or lag screws should not enter the plug. Forcing the screw shoulder into the plug adds nothing whatever to the holding power, and may bind or shear the screw—that is, turn the head off. Use a plug only as long as the threaded portion of the screw. If you are using sheet metal screws (threaded to the head), the plug can be the length of the screw minus the thickness of the work to be fastened. If the shoulder of your wood screw is longer than the thickness of the work to be fastened, countersink the plug in the masonry deep enough so that the shoulder will not enter.

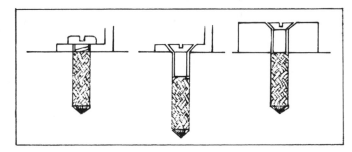

Proper length of a fiber plug.

When anchoring in hollow materials (plaster, hollow tile, Sheetrock, block, and so on), flare the end of the plug by turning a few threads of the screw into the plug in your hand. Push this assembly into the drilled hole and turn the screw in a few more threads. Then remove the screw, put the fixture in place, and turn the screw home.

Neoprene sleeves, which hold in a similar fashion as fiber plugs (except that they are made of neoprene rather than jute), are frequently used for mounting window fans, stereo speakers, and other high-vibration items. They are installed in the same manner as fiber plugs. Never use either fiber plugs or neoprene sleeves in masonry which is in poor condition. The holding power of these fasteners depends on the strength of the wall material.

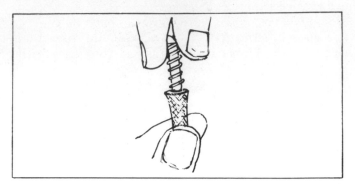

Flaring the end of a fiber plug.

Lag Screw Anchors These fasteners, also called expansion shields, are soft lead sleeves that can be attached with lag screws to concrete and other masonry surfaces. Specially designed horizontal fins prevent the anchor from turning in the hole as the lag screw is tightened, and the tapered annular rings add additional resistance to withdrawal from the hole. Because the fastener is threaded internally to match the threading of the lag screw with which it is used, it is fast and easy to install.

Lag screw anchors are grouped into two classes—short (1 to 2 inches long) and long (1½ to 3½ inches). Both take lag screws from ½ to ¾ inch in diameter. Short anchors are used to minimize drilling time in hard masonry; the long style is used in soft or weak masonry.

To install a lag screw anchor, drill a hole the same

Lag Screw Anchors—Size and Load Capacity		
Screw Size	Shield Length (inches)	Load (pounds)
¼	1, 1½	450, 600
5/16	1¼, 1¾	800, 1200
⅜	1¾, 2½	1200, 2000
7/16	2¼	1650
⅝	2, 3½	2500, 3500
¾	2, 3½	3000, 4000

size as the outside diameter of the shield. Set the anchor flush with or slightly below the surface. Insert the screw through the fixture and into the anchor. The anchor expands as the lag screw enters. The lag screw is driven with a wrench.

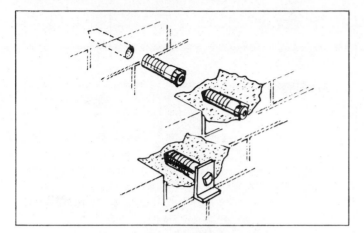

How a lag screw anchor is installed.

Lead Screw Anchors   These fasteners are similar to lag screw anchors except that they are used to fasten medium-weight objects with wood screws. Anchor lengths range from ¾ to 2 inches.

Select a screw length that equals the thickness of any mounting plate, plus the length of the anchor, plus ¼ inch. Drill the hole in the masonry surface the same size as the anchor and ¼ inch deeper than the anchor. Set the lead anchor flush with the wall surface, and drive the screw through the fixture into the anchor with a screwdriver.

Lead Screw Anchor—Size and Load Capacity		
Screw Size	Anchor Length (inches)	Load (pounds)
6-8 10-12	¾, 1, 1½	to 400
14	1, 1½	to 900
16-18 20	1, 1½	to 1300
22-24	1¾, 2	1600 up

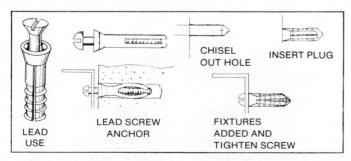

How a lead screw anchor is installed.

Machine Bolt Anchors   Machine bolt anchors are used to support an object that exerts a really tremendous pull—a large gate, for example—or that is subjected to great forces such as violent winds. There are two types of machine bolt anchors: caulking and noncaulking. The former is used if you cannot drill a hole of uniform diameter and perfect straightness. The latter, because of its length, is particularly good in weak masonry.

Machine Bolt Anchors—Size and Load Capacity			
Bolt Size	Shield Length (inches)	Hole Diameter (inches)	Load (pounds)
¼	1½	½	500
5/16	1½—2	5⁄8	800
5⁄8	1½—2⅜	¾	1000
7/16	2—2½	⅞	1000
½	2—2⅞	⅞	1500
5⁄8	2½—3¼	1—1⅛	2000
¾	3½—4	1¼—1⅜	2000
⅞	4—4¼	1⅜—1½	2300
1	4¼—4½	1⅜—1¾	2400
1¼	5½—6	2⅛	2500

The caulking anchor consists of a short sleeve wedged into a cap. To use the anchor, proceed as follows:

1. Drill a hole to the manufacturer's recommended diameter and depth. In weak or soft masonry, a slightly deeper hole may be drilled to countersink the anchor below the surface.

2. Insert the lead anchor into the hole with the threaded cone first, and then hit the caulk fastener with several sharp hammer blows. Some caulk-type anchors require a special setting tool to set the anchor in the hole.

3. Put the fixture in place and tighten the machine bolt. The minimum length for the ma-

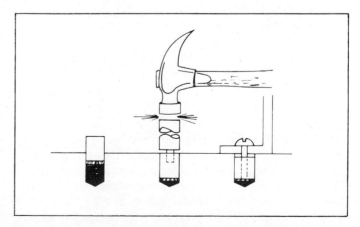

Installing a caulking-type machine bolt anchor.

chine bolt is one which will engage at least ⅔ of the threads in the cone after the fixture is in place.

The noncaulking anchor may be one of two styles: single and double. The former is considered the standard style of noncaulking machine bolt anchor and is made of rustproof zinc alloy. As the bolt is tightened, the threaded cone is drawn up into the expansion shield which develops a wedging action deep in the masonry where the masonry material is the strongest. One-piece construction permits fast, easy installation.

Two styles of noncaulking machine bolt anchors: (left) single and (right) double.

The double machine bolt anchor is designed for use in masonry materials of questionable strength or where heavy shear loads are encountered. Setting the anchor makes the opposing wedges at either end draw tightly into the anchor causing full-length expansion against the walls of the hole. The longer length of the double provides maximum gripping surface and will compensate for any voids or weak spots in the masonry. The top wedge-shaped cone acts as a bearing sleeve for the bolt when subjected to shear loads.

To install either style of noncaulking anchors, the following procedure should be taken:

1. Drill a hole to the manufacturer's recommended diameter and depth. Then, drop the anchor into the hole, making certain that the threaded cone is inserted first. For maximum expansion, the double style should protrude slightly.
2. Put the fixture in place, insert the bolt, and then tighten with a wrench until firm resistance to further turning is felt. *Note:* The bolt should be securely tightened to develop maximum wedging action.

Where it is desirable to set a double style anchor in an extra deep hole, it is recommended that a pipe sleeve of suitable size be set between the anchor and the fixture to be anchored. Deeper setting of the anchor will develop the maximum strength of the masonry material while the use of the pipe sleeve will provide additional support to the bolt if subjected to heavy shear stress.

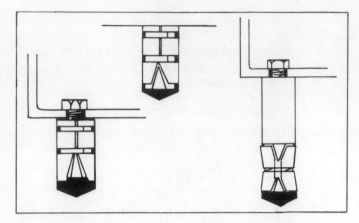

Installing a noncaulking type machine bolt anchor.

Stud Anchors Stud anchors are similar to caulking anchors in the way they are installed, but the anchor looks like a bolt without a head. When set into a masonry surface, the threaded portion of the anchor protrudes from the surface. The object which is being fastened is slipped over the threads and secured to the masonry with a nut. Stud anchors are generally available in sizes from ¼ by 1¾ inches to 1¼ by 12 inches.

The procedure for installing a stud is as follows:

1. With the fixture in place, drill a hole to the manufacturer's recommended depth. Maximum anchor performance is directly related to the accuracy of the drilled hole. An oversized hole will result in a marked reduction in the ultimate holding power of the stud anchor.
2. Insert the stud anchor in the hole to its full length and set it with several sharp hammer blows.
3. Tighten the nut with a wrench.

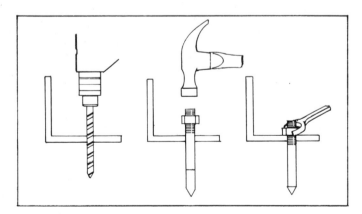

Installing a stud anchor.

Nail Drive Anchors Nail drive anchors, or pin rivets as they are sometimes called, are fast, easy-to-use, one-piece expansion anchors. Once the nail is driven home, the fixture is permanently locked in place, leaving it unaffected by vibration.

Nail Drive Anchors—Size and Load Capacity		
Shield Diameter	Shield Length (inches)	Load (pounds)
3/16	⅞	375
	1¼	110
¼	1	200
	1¼	240
	1½	325
5/16	1¼	300
	1¾	350
	2¼	375
	2¾	375
⅜	2	450
	3¼	485
½	2¼	450
	3½	485

To install a nail drive anchor, with the fixture in place, drill a hole the same diameter as the shield. Insert the anchor in the hole until the collar of the sleeve is flush with the fixture plate. Then, hammer the nail flush into the head of the sleeve. The anchor is now set.

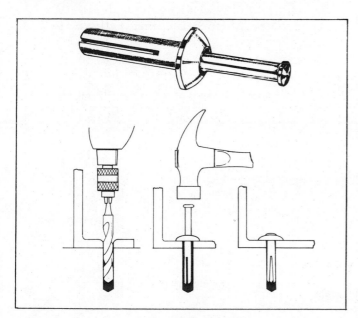

How a nail drive anchor is installed.

Another similar drive fastener is the one-piece expansion bolt. This fastener is made in four head styles: round, countersunk, stud-type, and tie wire. They are available in 3/16 to ½-inch diameters and in lengths of 1 to 6 inches. When driven into the proper diameter hole in hard masonry, the two sheared and pre-expanded halves of the drive expansion pin are compressed by the wall of the hole and will forever

try to regain their original bulged shape, thus exerting a tremendous force on the sides of the hole.

To install, drill a hole in the masonry the nominal diameter of the drive expansion bolt to be used and deeper than its length. Drill right through the fixture hole; no hole-spotting or positioning is required. The drive anchor must not touch the bottom of the hole; drill as deep as you like. Then, insert the drive bolt (leave the nut on the stud-type) and drive with a hammer, just as you would a nail. For round or countersunk fasteners, the anchoring job is now complete. If a stud driver is used, tighten the nut slightly with a wrench.

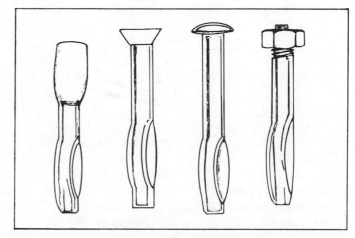

One-piece expansion bolt head styles: (left to right) tie wire, countersunk, round, and stud-type.

Because the resistance to withdrawal is so great with this type of anchor, it is advisable to use the stud-type in any case where the fixture might be moved at some time. If desired, the hole for the stud-drive may be drilled as deep as its overall length. Should the fixture be moved, the fastener may be buried by sending it all the way home.

Nail shields are also available for use in solid masonry walls. A hole the same diameter as the shield must first be drilled in the wall. The shield is then inserted, and the nail is hammered through the workpiece into the shield. This fastener gets its hold-

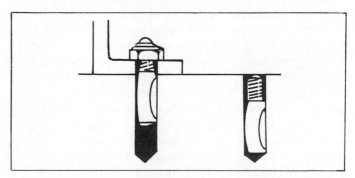

Burying a stud-driven type of bolt in the concrete.

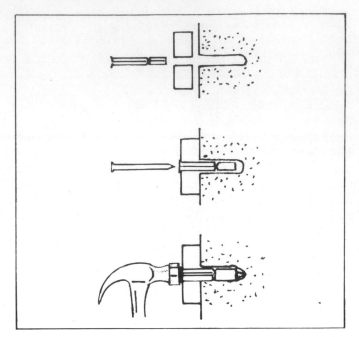

Installing a nail shield.

ing power from the shield rather than the nail, and it can support heavier loads than most designs of pin rivets but has nowhere the capacity of the one-piece expansion drive bolt.

Stud Fasteners   Masonry nails are strong but can be rather hard to drive, especially for the beginning handyman. To simplify this task, a tool called a stud driver has been especially designed to set special nails or stud fasteners in concrete, bricks, and even soft metals. The driver tool keeps the nails from buckling and holds them upright until they are driven home.

Stud fasteners from ¾ to 3 inches are usually available. When selecting the proper length, remember that the fastener should be long enough to penetrate from ½ to 1 inch; in masonry block and mortar joints from ¾ to 1¼ inches. Be sure to add the thickness of the material you are attaching to the amount of penetration needed for the best holding power. Most stud fasteners can hold loads up to 200 pounds each.

There are various types of stud fasteners. The nail-type is generally used to nail strips of lumber onto masonry surfaces. These can be driven directly through the lumber into the concrete. The threaded studs have a naillike point which is driven into the masonry surface so that the threaded end is left projecting. Nuts can then be set over the threaded end to hold the fixture in place. The washers that are attached to the fastener serve three functions: they aid in positioning the stud in the stud driver, they act as a guide to the stud while it is being driven into the masonry surface, and they help create a better grip after the object has been fastened in place.

Either type of stud fastener is placed in the holding sleeve or tube of the stud driver. An anvil in the other end of this sleeve projects at the top so that it can be struck with a hammer to force the pointed portion of the stud fastener into the masonry when the base of the driver is held snugly against the surface. While the stud driver can be used with a carpenter's hammer, it is best to use a 2- or 3-pound double-faced sledge, especially if you have many studs to drive. Of course, if you are driving a great many studs, it may pay to purchase or rent a power-actuated stud driver. This tool uses .22 cartridge loads to literally fire the stud fasteners into the concrete or masonry. Operational instructions for the velocity stud driver are in the applications manual which comes with the tool. Be sure to follow these instructions closely at all times.

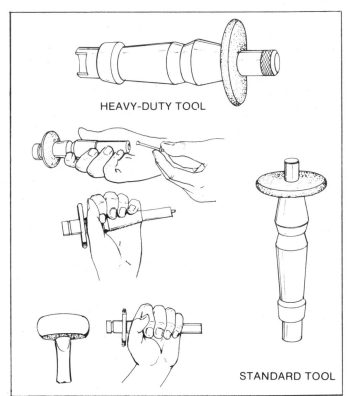

HEAVY-DUTY TOOL

STANDARD TOOL

A typical stud driver in use.

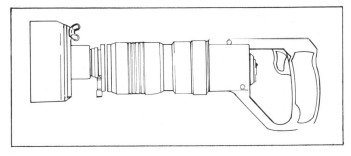

A velocity stud driver in use. In some communities, these power-actuated stud drivers can be rented from tool rental agencies.

Fastening with Adhesives and Glues

The ever increasing number of new basic adhesive compounds, together with all of the variations in each basic adhesive material, make the selection of a suitable formulation for a given application seem monumental at times. To those with minimal or practically no knowledge of adhesives and adhesion, the initial selection task could appear almost impossible. Fortunately, however, there are two major factors which, when taken into consideration, will greatly simplify the task. They are:

1. Type of material to be fastened. Are the materials to be joined porous or nonporous? Wood, cork, paper, leather, cardboard, and cloth are considered porous materials, while metal, glass, ceramics, porcelain, tile, and most plastics are nonporous. Are the parts easily clamped together? Are you trying to join two dissimilar materials such as wood and plastic? Are the materials rigid or flexible? Are the materials being bonded smooth or rough?

2. Physical conditions and characteristics. Will the completed bond be subjected to heat, cold, or moisture? Are there large voids that require a gap-filling function by the adhesive? To what kind of stress will the finished joint be subjected? Is the glue flammable? Are the materials structurally sound or are there internal weaknesses that will affect the strengths? What is the color of the adhesive when it is dried? Will any surface preparations be required to provide an adequate bonding surface? Is drying time important? Who is the end user of the bonded item, and where is the bonding taking place? If you are repairing a child's toy or working in an area with poor ventilation, you do not want to use an adhesive with a toxic effect.

With the answers to these important questions, you should be able to go a long way toward selecting the proper adhesive or glue for your fastening task. But, taking a look at the more popular adhesives and glues, let us attempt to clear up a point of confusion in the minds of some users: what is the difference between the terms adhesive and glue?

In today's world, the two terms are used almost interchangeably. At one time, the word glue referred only to products which were derived from an organic material and were used only on porous materials. The word adhesive was used only to describe products derived from a synthetic resin and used on nonporous surfaces. Nowadays, however, the majority of glues are made from synthetic resin bases, and most can be employed for other jobs in addition to their primary function of gluing wood and similar porous materials.

Types of Adhesives and Glues

Liquid Hide Glues The liquid hide or animal glues are actually a ready-mixed version of one of the oldest types of wood glue (if not the oldest) which furniture and cabinetmakers once cooked or boiled in a pot. Still made from animal hides, bones, and tendons, liquid hide glue is excellent for interior furniture construction and repair. It makes strong bonds and fills gaps in joints very well. It can be easily cleaned up with warm water.

Liquid hide glues set (harden) in two to three hours, but require at least eight hours at 70 degrees F. to completely dry. (Drying time and proper room temperature—hide glues tend to thicken as temperature drops below 70 degrees—can be critical under certain circumstances.) Bonding with this glue also requires clamping, but it does not creep under a load. The glue line dries to light brown or honey color. The major disadvantages of these glues are their poor water- or moisture-resistance and the importance of temperature control in their use. Incidentally, if you are a purist, flake or solid types of hide glue are available, but this material must first be soaked, then heated and mixed with a proper amount of water.

Resorcinols Resorcinols are completely waterproof, high-strength adhesives that are primarily intended for wood applications. They are excellent for outdoor furniture and for items immersed in water, such as boats (even toy boats). Resorcinols are a two-part adhesive, packaged in double cans. One can contains a cherry-colored liquid resin; the other contains a tan powdered hardener, or catalyst. The result of the mix is a dark brown, loose paste with a pot life of about two to three hours. When the mixture gets too thick to spread, it must be discarded.

When mixing a two-part resorcinol adhesive use matching measuring cups and spoons. This is to prevent any hardener from getting into the resin remaining in the original container by using the same utensils. Protect your eyes from both parts when mixing this type of adhesive. As powder has a tendency to pack down during storage, shake the closed powder can to fluff it and to make the measurement accurate. A stiffer mixture requires a shorter setting time than a wetter mixture.

Wood parts must be firmly clamped until the adhesive is completely dry. It cures in about ten to twelve hours at 70 degrees F., seven to nine hours at 80 degrees, and four to six hours at 90 degrees; but the clamps should not be removed for sixteen to twenty-four hours. It cleans off easily with warm water while it is still wet, but once it has hardened, it cannot be removed. Resorcinols have good gap-filling properties and do not creep under stress loads. They withstand freezing, boiling water, heat, fungus, and mild acids and alkalis. Their major disadvantages are that they must be used at temperatures above 70 degrees F., they have a relatively short pot life, and they require a long curing time.

Acrylic Resin Adhesive   This two-part adhesive (liquid and powder) provides an extremely strong bond that is waterproof and is not affected by gas or oil. Acrylic resin adhesives are good for filling gaps or cracks in objects that hold or are in water. It is most important that you follow the manufacturer's directions on the container to get the proper proportions for the job you are doing. Drying and setting time is controlled by the amounts mixed; for example, three parts of powder to one part of liquid will set in about five minutes at 70 degrees F. Changing these proportions will allow for faster or slower drying and setting time. Acetone can be used as a cleaning solvent. The glue line color of the dried adhesive is tan. Although it bonds to almost anything—wood, metal, glass, concrete, but not plastic—an acrylic-type adhesive sets too fast for large area work and is generally used for heavy-duty repairs.

Polyvinyl Acetate (PVA) or White Glue   One of the most commonly used adhesives, white glue can be used for all types of interior woodworking jobs, where waterproof joints are not required. It is also a good adhesive for paper, cloth, leather, and other porous materials.

White glue such as Elmers and Titebond is nontoxic and nonflammable. Clamping with moderate pressure is necessary, and workpieces should be kept at room temperature. While the adhesive sets in about one hour, full strength is obtained in approximately twenty-four hours. Surplus glue should be cleaned up with a damp cloth. When dry, it is translucent and can be sanded without clogging the sandpaper. A joint formed with polyvinyl white glue

will withstand only moderate stress. It should not be used on bare metal because it causes corrosion.

Polyvinyl chloride (PVC) is similar to polyvinyl acetate, except that it is highly resistant to moisture. It can be used on glass, porcelain, marble, metal, wood, and plastics. It dries fast (ten to thirty minutes) and clear. Its clean-up solvent is acetone.

Aliphatic Resin Adhesives   Growing more popular, this glue is cream-colored and works very much like white glue. It is, however, stronger than white glue. It has fast initial tack and sets in twenty to thirty minutes. It needs clamps for only half an hour and cures to full strength in twenty-four hours. Aliphatics have good heat resistance and are not affected by varnish, lacquer, and paint. They can be dyed or precolored with water-soluble dyes to match the material being repaired.

An aliphatic resin adhesive is water-resistant, but not waterproof. It is good for furniture repairs and can be used on wood, paper, leather, and other porous surfaces.

Casein Glues   An old reliable wood glue made from milk protein, casein glue comes as a light beige powder which you mix with water. The glue is not waterproof, but it is moisture-resistant and has long been used on exterior jobs. Casein glue can be applied at low temperatures (any temperature above freezing) and requires only moderate clamping pressure for two to three hours. However, the material may stain some dark or acid woods and tends to dull cutting tools when dry. It is excellent for use on oily woods, such as teak, lemonwood, and yew that will not take other kinds of wood glue.

Plastic Resin or Urea-Formaldehyde Glues   This light-colored, highly water-resistant wood glue is very strong but brittle if the joint fits poorly. It will not stain acid woods, such as oak and mahogany but should not be used with oily woods.

Plastic resin glue must be used at temperatures of 70 degrees F. or above. For maximum strength, joints must be smooth and fitted accurately. Firm clamping pressure must be maintained for at least twelve hours to insure a good bond. Plastic resin glue is resistant to rot and mold and will leave little or no glue line to mar the finished appearance. When mixed with wheat or rye flour and water in recommended proportions, it provides an easy way to glue veneers at relatively low cost. Plastic resin glues are odorless and nonflammable.

Cellulose or Plastic Cement   These glues dry clear and set quickly (in about ten minutes) with some pressure put on the joint, but the material takes about twenty-four hours to cure completely. They are waterproof, except on wood and some plastics, which they may also discolor. Most celluloses are flammable before they are dry. Their fumes are toxic and dangerous. Therefore, be sure to work in a well-ventilated room. Do not work with cellulose glues near a flame (including stove pilot lights or lighted cigarettes).

Cellulose cement is good for wood, metal, leather, paper, glass, and ceramics. It is also fine for jewelry repairs. These cements should not be used on objects that will be put under stress. Most brands are moderately water-resistant, but they all will weaken if the cemented joint is soaked for any length of time. Clamping is not required, but allow the cement to become slightly tacky before pressing the workpieces together. Hold them under hand pressure for a minute or two. (Some types, such as the ones used on model airplanes, give a good enough bond if the pieces are held together for fifteen to twenty seconds.) Clear plastic cement should not be used for repairing large-sized objects.

Cyanoacrylates In many ways, this is a remarkable adhesive. Better known as superglue because of its great strength, this group of adhesives sets as soon as the chemicals in the glue combine with moisture in the air. Surfaces bonded together with cyanoacrylate require no clamping and take hold within thirty to ninety seconds. Full strength is attained in twelve to twenty-four hours.

Cyanoacrylate glue bonds together practically all solid, nonporous materials either to themselves or to another nonporous surface. This includes metal, rubber, jewelry, china, glassware, most plastics, most ceramics, and some hardwoods. Cyanoacrylate adhesive should not be used on such porous materials as softwoods. (A few cyanoacrylates will adhere to softwoods; the addition of an ingredient called orthonol makes the difference.) Cyanoacrylate resists water, extreme temperature, and most chemicals. For effective storage, keep the container upright in a cool, dry location. Light, high temperature and humidity before it is applied to a surface reduce its effectiveness. Beware of getting this glue on your hands; it can join fingers so tightly that acetone nail polish remover may be needed to separate them. Acetone is the clean-up solvent. A cyanoacrylate glue is also a severe eye irritant.

Epoxy Adhesives This adhesive comes in two separate containers (usually tubes). One contains the epoxy resin; the other, a hardener. Read the instructions for the proper quantity to use. If both resin and hardener are not in proper proportion, the bond will be weak and may fail. Some epoxy containers automatically dispense the correct amount of resin and hardener through a tube, while others do not. An eyedropper is an accurate measuring device for small amounts of epoxy. Use small disposable cups for the actual mixing.

Once mixed, the adhesive remains in a workable condition for only a brief time. Therefore, try to mix only as much as you require and use the glue as quickly as possible. Clamp the work while the glue cures, which varies anywhere from five minutes to several hours. If you work in low temperatures (50 degrees F. or less), the epoxy will not harden. While acetone is considered the solvent, once an epoxy is applied, it is difficult to remove it from a surface.

When properly mixed, this adhesive provides a very strong bond but not a flexible one. Epoxy does not shrink when it hardens. It is waterproof and heat-resistant; repairs made with epoxy can be washed in a dishwasher after the adhesive has set. Although mainly used to bond nonporous material, such as glass, ceramics, metals, plastics, and china, it is also effective on wood and other porous materials. It is not used much on wood or absorbent materials, however, because other less-expensive glues are readily available for such materials. Re-

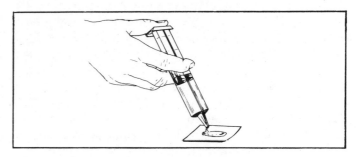

This dispenser solves the epoxy measuring problem by packing hardener and resin in rigid, transparent plastic tubes joined by a single nozzle. A double plunger at the top of both tubes squeezes resin and hardener through the nozzle in equal quantities.

cently, epoxy adhesives have been made available that come with a resin or hardener mixed with fillers to form a putty for various kinds of patching jobs, such as filling holes in gasoline and water tanks or sealing leaking pipe joints.

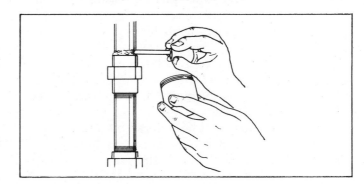

Epoxy is fine for patching a leaky water pipe.

Contact Cements Contact cement bonds dissimilar materials, such as wood, cork, leather, hardboard, and plastics. It can be used with metals, except brass, bronze, copper, or manganese. It is ideal for bonding large sheets of material to broad surfaces of wood, plywood, or even nonporous substrates. For example, decorative laminated plastics are generally bonded to the top of kitchen counters and bathroom vanities with contact cement. It is available with two bases: chlorinated and water. The latter is nontoxic and nonflammable.

This adhesive bonds on contact and requires no clamping. When applying, coat both surfaces generously and allow to dry before joining them together. This drying takes from ten to thirty minutes, but you must finish the installation in two hours. When both surfaces are brought together, the grip is so strong that the pieces cannot be pulled apart, so they must be aligned exactly before you join them. Be sure that the room temperature is at least 65 degrees F., or the bond will be weak.

Panel and Construction Adhesives   The application of plywood, gypsum board, and hardboard wall panels with panel adhesive is widely employed by homecraftsmen. Its use largely eliminates the need for brads or nails and the resulting concealment of their heads. Generally, the adhesive comes ready to use in a tube with a plastic nozzle. This tube fits into almost any caulking gun, and the panel adhesive comes out of the nozzle as a heavy bead. If the wall is in good condition, smooth and true, the adhesive can be applied directly to the back of the panel all around the edges in intermittent beads about 3 inches long and spaced about 3 inches apart. Keep the adhesive at least ¼ inch from the edges of the panel and be sure that it is continuous at the corners and around openings for electrical outlets and switches. Additional adhesive should be applied to the back of the panel in horizontal lines of intermittent beads spaced approximately 16 inches apart. Once the adhesive is applied, the panel may be pressed against the wall. It may be moved as much as is required for satisfactory adjustment. To make this easier, drive three or four small finishing nails about half their length through the panel near the top edge. The panel can then be pulled away from the wall at the bottom with the nails acting as a hinge. After any adjustment has been made, a paddle block should be used to keep the panel pressed back on the wall, and then the nails are driven home. (These will be covered by a molding.) A rubber mallet or a hammer and padded block should be used on the face of the panel to assure good adhesion between the panel and wall.

This adhesive also may be used on furring strips and open studs. It is applied directly to each furring strip or stud in continuous or intermittent beads. Panels are then applied by the same method as just described above. Never apply adhesives on plaster walls in poor condition, with flaking paint or wallpaper that is not tightly glued. If the plaster seems hard and firm and does not crumble when you drive a nail into it, it is probably safe for adhesives.

In addition, panel adhesives and newer construction adhesives are used in the application of subfloors, insulation, sheathing boards, and polystyrene or plastic foam panels. The use of these adhesives reduces nailing and saves a great deal of time. Special metal framing adhesive was designed to meet the growing demands of galvanized steel and aluminum framing members both in floor and wall systems. Still another similar material is concrete adhesive. This ready-to-use liquid permanently bonds new concrete to old, as well as new concrete, plaster, or stucco directly to many surfaces, including wood, hardboard, brick, ceramic tile, wallboard, concrete block, metal, and others.

There are other adhesives used in home construction and remodeling work, including floor tile adhesive, wood parquet floor mastic, and carpet adhesive.

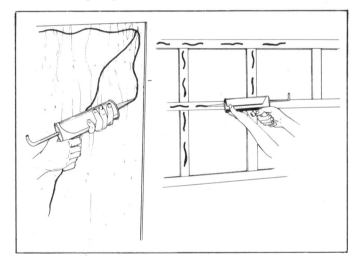

Applying a panel adhesive to the back of a panel and furring strips.

Hot Melt Glue   This is a waterproof, fairly strong glue that comes in stick form and is applied with an electric gun. It does not require clamping because it sets quickly—in about sixty seconds. Hot melts will bond most porous materials. No toxics are given off and the adhesive is not flammable.

While various glue guns have different features—preset thermostat, aluminum melting chamber, self-standing design, two-cartridge storage, and so on—the following basic instructions hold for all models.

1. Insert the glue stick cartridge in the gun, in the proper chamber as detailed by the manufacturer's instructions. Some guns require the insertion of two cartridges when first used, because they feature a reserve glue compartment.

2. Plug the gun into a 120-volt electric circuit (unless it is designated as a 240-volt unit), and allow three minutes for warm-up. Take care not to touch the nozzle or heating chamber. The temperature in this area may reach as high as 400 degrees F.

3. When the three-minute warm-up period has passed, squeeze the trigger of the tool to dispense the glue. From the time you start applying the adhesive to the surface to be bonded together, you have about ten seconds of time. Work quickly. Do not try to cover a large area. Place the two surfaces to be bonded together within ten seconds. As with any adhesive, the two bonding surfaces must be clean.

4. After sixty seconds the adhesive will attain 90 percent of its bonding ability, which is sufficient to hold the surfaces together.

5. When you are finished with the glue gun, pull the line plug and rest the tool on a nonflammable material or hang it on a hook to cool. Allow the glue to remain in the tool; it can be reheated repeatedly without affecting its bonding ability. Never change the nozzle when it is hot.

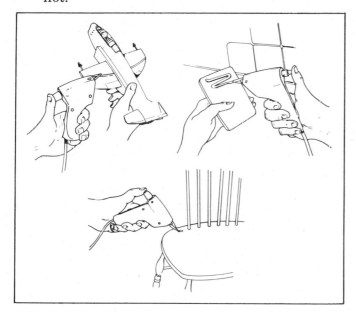

Applying hot melt glue with an electric glue gun.

Other Household Glues There are other household glues that are available. For instance, library paste, made of white flour and water, is good for fastening paper and is used for mounting artworks. Mucilage is another good paper bonder.

Rubber cement is a fast-drying adhesive that can be used to bond cloth, leather, and rubber. Apply the cement to one or both surfaces to be bonded, then join. If the cement thickens, add a small amount of rubber cement thinner to it and stir thoroughly.

There is a whole group of acrylonitriles and fabric-mending adhesives available that will repair tears in canvas, felt, wool, cotton, and other fabrics, as well as in leather and leatherette. They can be employed with or without a supporting patch, and in most cases, the bonds are stronger than stitching. All will stand normal washing and ironing, but some brands will not stand up under dry cleaning fluids. Acrylonitrile adhesives bond a variety of materials, including glass and metal. They are not, however, recommended for woods.

Silicone glue is a flexible, waterproof, rubberlike cement. It works well on porous and nonporous surfaces, such as glass, metal, fabric, porcelain, and pottery. It also can be used as a sealer around shower and tub tile.

Styrene-butadiene adhesives are good for replacing loose tiles and bricks, as well as attaching kitchen fixtures (tile and nontile) and other nonporous surfaces. It has excellent gap-filling properties that provide a grip on irregular surfaces. This adhesive can be drilled, sanded, and painted when hard. It should be used where the adhesive will be exposed to oil or gasoline.

Buna-N base adhesives are made from synthetic rubber, and they are used for bonding two materials with a flexible joint between them. They will bond almost anything, allowing ample back-and-forth movement. When the adhesive is employed on porous materials, the moving parts should be assembled while the adhesive is wet. However, when making a bonded flexible joint between two nonporous materials, apply the adhesive, let it dry (about two to four hours), then soften it with heat just before bringing the parts together.

There are several adhesives on the market that are designed for repairing rigid and flexible plastic items that are made of such materials as acrylic, styrene, vinyl, and phenolic resin. This adhesive dries clear and is fairly flexible, thus it can be used to mend such things as torn plastic raincoats, blow-up toys, air mattresses, and above-ground swimming pools.

Metal bonding adhesives are available for patching cracks and for bonding two small metal parts together. Since these adhesives have special patented formulations, it is difficult to describe how they are applied. Just be sure to follow the directions on the container for each type of metal-bonding adhesive.

Actually, it is very important to always read the manufacturer's recommendations printed on the container for the use of any glue or adhesive. These instructions will state the type of materials that can be bonded, the length of drying time, the solvent or cleaning agent, application, and whether or not pressure is required on the pieces being joined. All labels state very clearly any hazard which may be encountered and every precaution that should be taken. In other words, if you want to obtain the best possible results, these instructions should be observed to the letter.

Tips on Applying Adhesives and Glues

Knowing what adhesive to use is only part of the job; you must know how to apply the adhesive as well. While the instructions on the container will give specific directions for a particular adhesive, there are certain general tips that hold good for most gluing jobs. They are as follows:

1. If you are not sure of the bonding ability of the materials and adhesive, test a small area first. Sometimes a fast-setting adhesive might set up too quickly if you are assembling many pieces at the same time. A slower-drying adhesive would be preferable for this type of job.

2. Temperature is important. You will get better results if the work, adhesive, and room are at the temperatures recommended by the manufacturer. Keep in mind that when lumber and other materials are brought in from the cold outside, it takes several hours before they warm up to room temperature.

3. Pre-fit all parts to avoid costly errors. A test assembly should be made without glue, as a trial run, before actual pieces are fastened together. This will enable you to adjust your clamps to the correct settings and cut down on assembly time. This is important, especially if you are using a fast-setting adhesive.

4. Hard, smooth wood surfaces will glue better if the wood is roughened slightly by lightly sanding across the grain. Roughing the surface of some metals is also recommended.

5. When gluing metal, soak it in acetone, then let it dry. Do not touch the clean metal area before applying the adhesive.

6. Both surfaces to be joined must be clean before any adhesive is applied. When making furniture repairs, be sure to scrape off all traces of the old finish or glue before assembling, so that the adhesive can penetrate into the wood to insure a good bond.

7. Joints should be fitted as tightly as possible. Whereas many glues have good gap-filling properties, all adhesives hold better and joints look better if the pieces are snugly fitted.

8. The end grain of most wood is highly absorbent, and if it is permitted to soak up the mois-

ture of the adhesive, a weak joint may result. To prevent this, give the end grain a thin prime coat of the adhesive you are using, applying it a few minutes ahead of time. Be sure to give the end grain a second coat of the adhesive when gluing other parts of the joint.

9. Apply the adhesive evenly in a smooth coat over all areas of contact. Wherever possible, apply the adhesive to both surfaces. This will insure uniform coverage.

10. Do not over-glue; use just the amount needed. To obtain uniform coverage, choose the applicator according to the size of materials to be glued and the type of adhesive. Q-tip sticks and wooden tongue depressors are readily available for small projects. Brushes permit full and uniform coverage of the adhesive, while trowels, rollers, or the serrated edge of a hacksaw blade can be used to spread adhesive on broad surfaces.

11. Be sure that the work is firmly clamped in position before starting the drying period. Employ just enough pressure to close the joint tightly, but not too much to warp or twist the pieces out of alignment. Also, too much pressure may squeeze out so much adhesive that

Chair rungs and similar joints can be reglued even when it is not feasible to pull them apart. Drill a small hole into the joint and inject the glue with an oil can.

the joint will be starved. Remember to clean away excess glue immediately, as some adhesives stain.

12. No matter what type of adhesive is used, give it plenty of drying time. An ordinary heat lamp will speed up the drying of the joint.

Clamps and Clamping

Some adhesives require clamps to maintain a constant, firm pressure between surfaces being

Common Repair Jobs and Their Adhesives

Key:*

A—Liquid Hide Glue
B—Resorcinol
C—Acrylic Resin Adhesive
D—Polyvinyl Acetate (PVA) or White Glue
E—Polyvinyl Chloride (PVC)
F—Aliphatic Resin Adhesive
G—Casein Glue
H—Plastic Resin or Urea-Formaldehyde Glue
I—Cellulose or Plastic Cement
J—Cyanoacrylate
K—Epoxy
L—Contact Cement
M—Panel and Construction Adhesive
N—Hot Melt Glue
O—Metal Mender

Repair Task	A	B	C	D	E	F	G	H	I	J	K	L	M	N	O
General woodworking, light duty, interior	X			X									X		
Heavy-duty woodworking		X	X				X								
Waterproof wood joints		X	X	X		X						X			
Laminating plastics to wood or metal					X						X				
Gluing metal to wood			X		X						X				
Repairing and bonding metals			X		X				X		X				X
Ceramic and masonry repair			X						X		X				
Bonding glass to glass or metal			X		X				X	X	X				
Bonding plastic to metal, wood, plastic					X				X	X	X				
Repairing flexible plastic					X										
Mending leather, cloth, canvas				X		X									
Leather to wood or metal					X				X			X			
Rubber to metal, wood, glass			X		X						X				
Installing hardboard, plywood, gypsum board panels				X						X		X	X		
Cloth to wood, plaster, cardboard				X	X										
Cardboard or paper to itself or to wood				X	X										
Porcelain and china, light duty			X		X				X		X				
Porcelain and china, heavy duty									X	X	X				
Exterior oily woods						X									
Veneer, unoily woods							X								

*Miscellaneous glues such as library paste, mucilage. rubber cement, acrylonitriles, fabric mending, silicone, styrene-butadiene, and buna-N base are not included in this chart.

glued until the adhesive sets up. There are many different types of clamps, but virtually all are used for the same purpose—to hold two pieces of work together for convenience or during the drying of an adhesive. Among the clamps which the craftsman might use in the shop are the following:

C Clamps Most commonly used in the workshop is the C clamp, shaped like the letter C. It consists of a steel frame threaded to receive an operating screw with a swivel head. C clamps are made for light, medium, and heavy service with a variety of openings from ⅝ to 12 inches. Protect the work by inserting small blocks of wood under the jaws of the clamps.

Hand-Screw Clamps These clamps consist of two hard maple jaws connected with two operating screws. They are a generally preferred holding device for nearly all types of shop projects and repair work. They grip and hold odd shapes securely and will not mar highly finished surfaces.

To use hand-screw clamps correctly, learn the habit of grasping the end spindle with the right hand; then the direction for swinging or rotating the hand screw to open the jaws will always be the same. Rapid adjustment of the hand screw is obtained by proper swinging. Hold the handles firmly, arms extended, and, with a motion of the wrists only, make the jaws revolve around the spindles. When the jaw opening is approximately correct, place the hand screw on the work with the end spindle either to your right or in the upper position, and with the middle spindle as close to the work as possible. Adjust either or both handles so that the jaws grip the work easily and are slightly more open at the end. Turn the end spindle clockwise to close the jaws onto the work. Final pressure is applied only by means of the end spindle. The middle spindle acts as a fulcrum. Make certain that pressure is applied all along the entire length of the jaws, not just at the end or at the edge of the work.

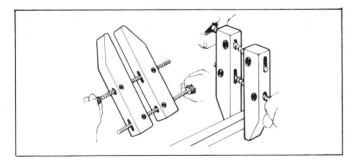

Proper use of spindle hand-screw clamps.

Bar Clamps Bar clamps are long (the standard maximum opening ranges from 2 to 8 feet) precision clamps normally used for gluing fine furniture. They

adjust by means of a movable jaw at one end of a metal or wooden bar and a crank that turns a screw at the other end. Scrap wood pads must be employed under their jaws to prevent marring.

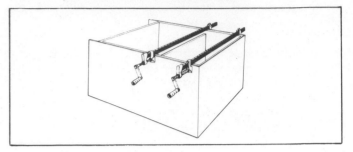

Metal bar clamp in use.

Pipe Clamps Pipe clamps operate in the same manner as bar clamps except that they slide on pipe. These clamps are available to fit either ½- or ¾-inch-diameter iron pipe. Only one end of the pipe need be threaded. The craftsman should have several different lengths of pipe to use with the clamps. While you can use a long pipe for all jobs, the excess pipe might get in your way. The pipe, of course, must be straight and smooth.

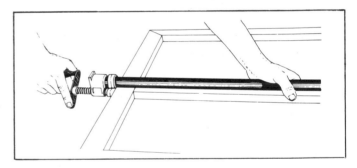

Typical pipe clamp.

Spring Clamps Resembling large clothespins, these handy clamps are available in sizes that open from 1 to 4 inches. They are the extra hands around the shop that hold your work in position while you are busy with gluing another part of the job. They can be used where light pressure is adequate. Some spring clamps have plastic-covered tips to minimize marring of the work.

Band Clamps These clamps solve the knotty problem of clamping round or irregular shapes where uniform pressure is required simultaneously at several joints. They are especially efficient for clamping furniture. The canvas or steel band encircles the work and is pulled tight from either end through a screw-clamp device. The self-lock cams of the clamp hold the band securely without slippage while final screw pressure is applied. Slight pressure on the cam extensions releases the band instantly.

The canvas band, which is usually about 2 inches wide, is recommended for most applications.

Web clamps are lightweight, low-priced band clamps with innumerable uses. The 1-inch-wide nylon band (which is usually 12 to 15 feet long) can be placed around any regular or irregular shape to apply clamping pressure all around the work—drum tables, chair frames, picture frames, and so on. With the band so located, it is drawn snug by hand, and final pressure is applied by means of a wrench or screwdriver applied to the hexagonhead slotted tightening bolt.

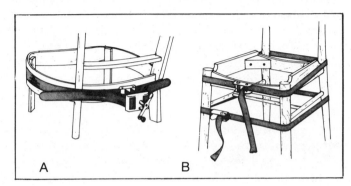

Band clamp (A) and web clamp (B) in use.

Three-way Edging Clamps Three-way edging clamps provide a convenient, practical method for applying right-angle pressure to the edge or side of the work. A unique three-screw design permits the right-angle screw to be centered or positioned above or below center on varying thicknesses of work.

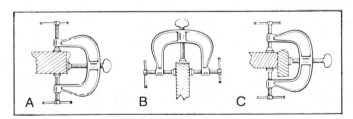

Three-way edging clamp and how it may be applied (A) with the right-angle screw off center, (B) centered, or (C) to clamp around returns.

Miter Clamps There are several types of miter clamps available. They are used in picture frame making. The miter clamps are designed for mitering flat casing where a ⅝-inch-diameter blind hole can be bored in the back of each piece. It forces the two ends being mitered directly against each other, no matter what the angle, with no possibility for the ends to creep along or away from each other.

The corner clamp is a flat triangular device used to clamp the corners of furniture frames, picture frames, and the like. The frame is set into the clamp, and a diagonal bolt then pushes it against the cor-

ner. In a variation of this device, pressure is exerted by two bolts which tighten against the outside corners of the frame.

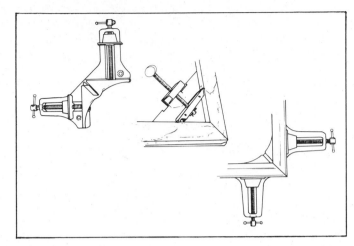

(Top) Typical miter clamp; (center) typical corner clamp and how it holds; (bottom) typical miter overall clamp.

Bench Hold-Down Clamps These hold-down clamps can be installed in wood or metal surfaces. This is a distinct advantage over conventional clamps and vises which function only near the edge of a workbench or table. Without the use of tools of any kind, the clamp slides onto a prespotted holding bolt, ready to go to work instantly. That is, it will hold work in the middle or near the edge of metal or wood benches, T slot or machine tables. Pressure is applied to the work by a screw adjustment. The clamp slides off the bolt head to be stored when not in use. When the hole is countersunk, the bolt head drops below the surface, leaving the work area clear and uncluttered. The clamp swivels 360 degrees around the holding bolt to secure the work in position at just the right spot or angle. One clamp, with several bolts installed at strategic locations, will perform work-holding service in many places throughout the shop. When the device is installed along the side of a bench for use as a vise, a cotter pin (fur-

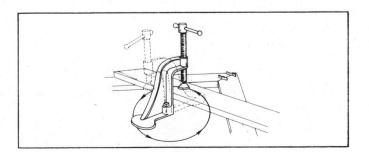

Bench hold-down clamp in use.

nished) run through holes in the clamp base holds the clamp on the bolt head to keep it in place when the screw pressure is released. Press screws are similar and are used frequently in the making of a veneer press.

Edge Clamps These handy attachments provide pressure at right angles to the axis of the bar of the clamp used. They are used for applying moldings to edges, drawing joints together, or for applying pressure to the middle of board areas.

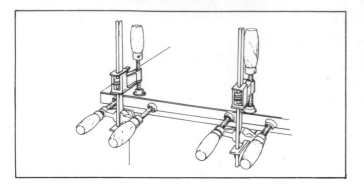

Edge clamp fixtures in use.

General Uses of Clamps When using clamps, first consider your clamping requirements, and then select the clamp best suited to your needs by ascertaining: (1) the opening required; (2) the depth required; (3) the strength and weight required; (4) whether or not a full-length screw is essential (if not, the constant hindrance of the screw extending beyond the frame can be eliminated by selecting a clamp with a screw length proportionate to your needs); (5) the type of handle best suited to your needs; and (6) the balance of clamp operating time versus clamping needs, i.e., do you require a spring clamp, C clamp, bar clamp, or some other type. This appraisal of your clamping needs will assure you of getting the most from your clamps by saving time and money and enhancing the life of the clamps.

With all clamps (except possibly the hand-screw type), it is important to remember that heavy pressure crushes softwoods. To keep from ruining the workpiece, large blocks are used to distribute the pressure over a wider surface. Provide yourself with a stock of blocks of assorted sizes, thicknesses, and lengths, preferably of hardwood, such as maple or birch.

Use clamps in pairs on both ends of the work. This will prevent one end from separating while the other is being joined. For large surfaces, additional clamps are needed.

Apply even pressure on all clamps. Tighten as far as possible with the handle. After a few minutes, take a few extra turns on the clamp handle, if possible. Avoid pressing in the sides of the workpiece. When closing the jaws of a clamp, avoid getting any portion of your hands or body between the jaws or between one jaw and the work.

Other Ways of Holding Glued Parts

When a clamp is not available, there are alternative holding devices that can be made to exert pressure on glued parts. One of the simplest is the rope tourniquet. Wrap a length of rope around the pieces that must be held together, and apply pressure as needed by using a short length of wood to twist the rope in the center. Wedge the stick in place. Be sure to protect the surface from rope damage by inserting folded pads of heavy paper or cardboard at all pressure points.

A stack of books can weight down small surface areas while a glued joint dries. For larger horizontal areas, use a pile of heavier weights, such as concrete blocks, bricks, or buckets of sand. A protective sheet of plywood should be laid over the work first to prevent marring.

When gluing small surfaces that are difficult to clamp, cellophane tape will hold the pieces together while the glue dries. Awkward joints (a cup handle, for instance) can frequently be held by two sackfuls of beads, marbles, or lead shot, tied together and slung over the surfaces to be held.

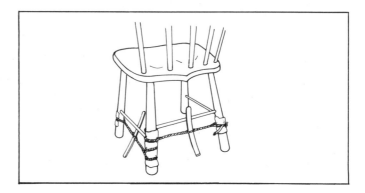

A rope tourniquet in use.

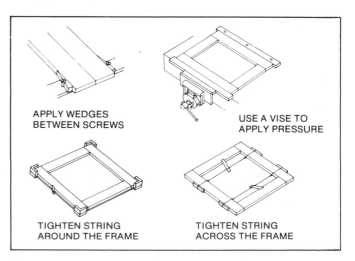

APPLY WEDGES BETWEEN SCREWS

USE A VISE TO APPLY PRESSURE

TIGHTEN STRING AROUND THE FRAME

TIGHTEN STRING ACROSS THE FRAME

Several homemade clamping devices.

Joining Materials

Applying the right adhesive correctly will create a bond that is stronger than the materials being joined. The application of the adhesive itself is easy, but some projects can be made simpler by combining the following information with the manufacturer's directions.

Wood-to-Wood There are three basic types of wood gluing and assembly: edge-to-edge, face-to-face, and permanent joint gluing. These are used in woodworking and in building construction.

Edge-to-edge gluing increases the width of the surface. The adhesive is lightly spread by brush or applicator on the edges being joined. When gluing edge to edge, the grain of the wood should be alternated to prevent warping. Bar clamps are used to hold the pieces together with enough pressure exerted to squeeze out a small amount of glue around the glue line.

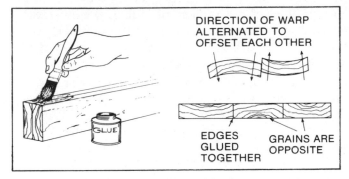

Edge-to-edge gluing.

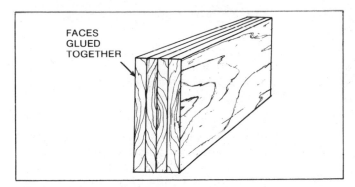

Face-to-face gluing.

Face-to-face gluing is required when the thickness of the stock has to be built-up or increased. A thin, even coat of adhesive should be applied over the surfaces to be joined. Hand-screw clamps should be carefully used to apply pressure equally, avoiding uneven adhesion of the surfaces.

The objective of permanent joint gluing is to hold the joint in place, which in turn provides strength and rigidity to the object. Clamps are usually re-

Gluing Properties of Hardwoods and Softwoods	
Hardwoods	
Group 1 (Glue very easily with different glues under wide range of gluing conditions)	Group 2 (Glue well with different glues under a moderately wide range of gluing conditions)

Aspen
Chestnut, American
Cottonwood
Willow, black
Yellow poplar
	Alder, red
	Basswood
	Butternut
	Elm:
	    American
	    rock
	Hackberry
	Magnolia
	Mahogany
	Sweet gum

Group 3 (Glue satisfactorily under well-controlled gluing conditions)	Group 4 (Require very close control of gluing conditions or special treatment to obtain best results)

Ash, white
Cherry, black
Dogwood
Maple, soft
Oak:
  red
  white
Pecan
Sycamore
Tupelo:
  black
  water
Walnut, black
	Beech, American
	Birch:
	  sweet
	  yellow
	Hickory
	Maple, hard
	Osage orange
	Persimmon

Softwoods	

Bald cypress
Cedar:
  Alaska
  eastern red
  western red
Fir, white
Larch, western
Redwood
Spruce, Sitka
	Douglas fir
	Hemlock, western
	Pine:
	  eastern white
	  ponderosa
	  southern yellow

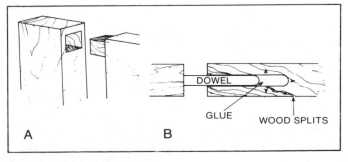

Permanent glue joints.

quired after a thin coat of adhesive has been applied to the joint area.

In woodworking, when using dowels, apply the adhesive to the dowel and not into the hole. Placing the adhesive in the hole and then forcing the dowel into place could cause the wood to split and break.

Plastic Laminate-to-Wood When applying plastic laminate to wood, always cut the plastic about ¼ inch greater in length and width than the surface to which it will be bonded. If you are using a portable saber saw, remember to mark and saw the panel with the decorative face side down, since the blade cuts on the up stroke. With a table saw, the material is cut with the finished side face up. For a neat job, regardless of what tool was used, trimming is always done after bonding.

Generally, it is best to apply the cement to the back side of the laminate first, then the core piece. Since the latter is more absorptive, glue dries faster on it. Cement may be applied with a paintbrush, however, a short-nap paint roller does a better job of distribution and takes less time.

The order in which to apply the various panels is determined by the edge most visible to the eye and most subject to abuse. Thus, self-edge strips always go on before the top panel.

Once you have decided upon sequence, start by applying the contact cement to the core stock surface and its mating piece of laminate. Check the instructions on the label for drying under the right climatic conditions. The parts should be ready for bonding in about thirty minutes. To check, touch the surface in several places with clean kraft paper; when the adhesive no longer adheres to the paper, the surfaces are ready for bonding.

Keep the mating pieces apart until they are ready for contact. Fingers alone can guide small pieces, but when bonding larger pieces, a helper will be needed to prevent accidental contact. There are several ways to keep the pieces apart until you are ready to bond them. One is to use a large sheet of clean kraft paper as a slip sheet; another is to use clean ¾-inch diameter dowels or square sticks to keep the laminate and core stock separated. Align the laminate over paper or sticks; check all four sides to make sure the core stock will be covered; then, starting at one end, remove the paper or sticks and bond the laminate to the core stock.

Most contact cements need only momentary pressure after bonding, but do not mistake momentary for light. Immediately upon bonding the laminate to the core stock, apply as much pressure as you can over the entire surface. A J-roller is an inexpensive tool and is preferred to assure adequate pressure. Even a rolling pin may be used. You can also slide a

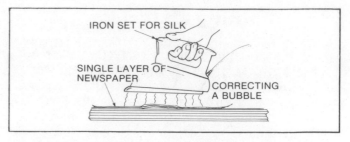

If a bubble should form, place an iron on it and press down until the heat penetrates the area. If the newspaper scorches, lower the heat setting. Roll out the blister.

block of clean wood about the surface and give it hefty blows with your hammer. Do not worry about giving the laminate too much pressure; you cannot. Just make certain you are careful when applying pressure near overhanging edges. If a bubble should form, it can be removed with an iron.

The final step is trimming. The easiest method, of course, is with a router and carbide cutter. Lacking this power tool, you will have to trim the overhang using a block plane and smooth file. Either way, the overhang is first trimmed flush (90 degrees) and then beveled slightly to about 20 to 22 degrees. Finish the beveling work with a smooth file.

To remove contact cement remnants from the surface, use the contact cement solvent recommended on the label instructions, and use scraps of plastic laminate—never use metal—to scrape off the heavy globs. Be sure to follow safety precautions for the solvent's use and provide adequate ventilation.

Metal-to-Wood Relatively new in the field of woodworking are the glues which produce high-strength durable metal-to-wood bonds. Many of these glues combine a thermosetting resin (frequently of the phenolic type) and a thermoplastic resin or elastomer. The latter is usually a polyvinyl resin, synthetic rubber, or epoxy. The vehicle may be water, alcohol, ethyl acetate, or another solvent.

The application of metal-to-wood adhesives may be accomplished in one of two ways. The first requires that the adhesive be spread on one or both surfaces to be joined. The solvent is then evaporated and the bonding completed under heat and pressure. The second method involves two steps and usually two adhesives. To begin the procedure, the metal is primed with the first adhesive and cured at elevated temperatures. Next, a woodworking type of resin adhesive that sets at room temperature is applied to the already primed metal. Finally, the metal surface is thoroughly cleaned with acetone, using a great deal more care removing solvents before bonding than when joining wood to wood. Always be sure to carefully read the manufacturer's instructions when making a metal-to-wood bond.

Fastening Wood Joints Together

To insure strong, rigid joints in all types of wood-working projects, there are a number of different wood-joining techniques that can be employed. Some of these techniques have already been fully described in previous chapters of this book. Nails, screws, and adhesives are all used to put wood joints together. In fact nails and screws often combine with a wood adhesive to make a very strong, rigid joint.

The type of wood joint has a great deal to do with its strength and rigidity. While these are easy to make with power tools, they can be made with common hand tools as well. When these joints are reinforced with an adhesive and/or screws or nails, they are stronger and more rigid than the simple butt joint; but even this joint can be strengthened by dowels or reinforcing corner blocks.

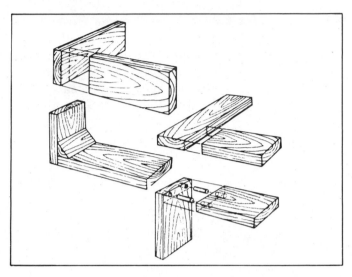

Butt joints and ways to strengthen them. For the reinforcement to work, however, the edges to be joined must be tested for absolute squareness before the pieces are fitted together.

Dowels and Splines

If an exceptionally strong joint is required, dowels and splines are preferred over nails or countersunk screws. Dowels are, of course, hardwoods—generally birch or maple. They are available in diameters from ⅛ to 3 inches, with either a plain or grooved surface. The latter surface allows the glue to run more freely into the joint. The grooves are generally cut on a band saw or lathe.

When selecting the size of a dowel rod to use, a general rule is that the diameter should be no more than half the thickness of the stock. The depth of the hole will vary with the type of joint. Remember that a dowel inserted in the drilled holes of a joint adds considerably to the strength of the joint. The dowel's hardness and the fact that it is used with the grain at right angles to the materials joined are the reasons for this added strength.

Actually, there are two methods used in doweling, the open method and the blind method. In the open method, a hole is drilled completely through one piece of wood and deeply into or through the piece to be joined. The dowel is coated with glue and pushed completely through the drilled holes, joining the pieces. The remainder is then sawed off flush with

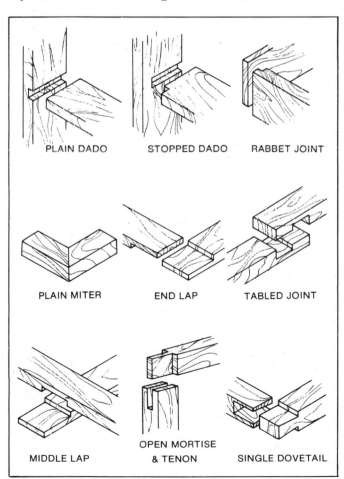

PLAIN DADO	STOPPED DADO	RABBET JOINT
PLAIN MITER	END LAP	TABLED JOINT
MIDDLE LAP	OPEN MORTISE & TENON	SINGLE DOVETAIL

Common wood joints.

the outer surface. Dowel stock for open doweling is kept long to allow for a flush cut after the joint is made.

In the blind method, holes are drilled partway into each piece from the joined faces. A rule of thumb is to drill the holes in each piece to a depth of approximately four times the diameter of the dowel. A dowel is then glue-coated and inserted in one hole, and the second piece is pressed onto the protruding dowel end. The length of the dowel rod should always be cut about ¼ inch shorter than the total of

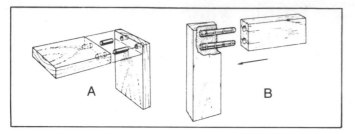

(A) An open dowel joint; (B) a blind dowel joint.

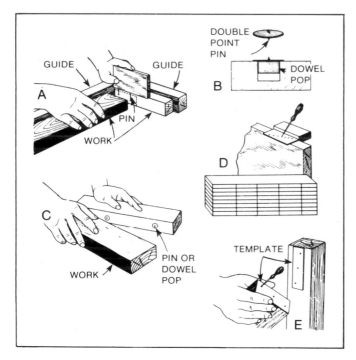

Methods of laying out dowel holes.

the two holes. Cut the ends of the dowel pins with a bevel. To ensure adequate glue on a snug-fitting dowel, some are spiral-grooved and others are provided with a lengthwise slot. When filled with glue, these indentations assure a firm grip. In most joints two or more dowels are used instead of one to prevent the pieces from twisting on the dowel.

Various methods are used for laying out the hole positions. Ordinary pins are stuck in a block of wood and placed in such a position as to come between joining members, as shown. When the joining mem-

bers are pushed forcibly together, the pinheads make an impression on each piece, which serves as a guide when drilling the dowel holes.

Double-point thumbtacks can be used in the same manner. The more standard method of working, however, is to use dowel centers or "pops." Here, after drilling the holes for the dowels in one piece of wood, you insert dowel centers in these holes. Then, you align the two pieces of wood as they will be joined. When you press them together, the points on the dowel centers mark the second piece of wood. It is now possible to drill holes at these center marks. When the pieces are connected with dowels, the blind dowel joint is perfectly aligned. Dowel centers come in assorted sizes to fit holes from ⅛ to 1 inch in diameter.

When locating dowel holes in a series of boards that will be joined edge-to-edge, position the board edges and butt them surface to surface. Using a combination square, mark the hole location on one edge and carry the line across all the pieces. Identify the board faces which will be at the top after assembly. Drill dowel holes in the edges at the crosslines, using a dowel jig or drill guide that will gauge holes automatically.

This doweling jig has a revolving turret that moves on parallel arms. Its capacity is up to 4 inches, and it provides six hole guide sizes from 3/16 to 1/2. You can clamp it to a single board as shown or to several when you wish to match holes.

When doweling joints other than those at edges, it is a good idea to make a template of stiff cardboard, thin plywood, hardboard, or even sheet metal if the long-term use justifies it. Drill 1/16-inch or smaller dowel center holes at the desired locations. Locate the template accurately first on one piece, then the other, to mark dowel centers with a center punch or awl.

When using the open method of doweling, the problem of aligning drilled holes does not occur.

Even when poorly centered, the holes match and the dowel can be driven through both holes. In the blind system, however, two separate holes must be drilled, and here trouble can develop unless a jig is used.

In some instances, the drilling device is clamped into position, and the wood is guided to it along measured channels to ensure proper centering of the holes. In others, the wood is clamped, and the drilling device is guided along identical channels for all holes drilled. In either case, the guide is actually a jig. Furthermore, it is also possible to arrange the jig so as to control the depth of the hole drilled.

To further insure that dowels do not turn or become loose, thereby permitting the joint to crack or spread as the surrounding wood shrinks, small finishing nails or brads may be set through wood surfaces into the dowels.

Splines are used to strengthen all types of joints from plain butt to fancy miters. The spline itself is a thin strip of hardwood or plywood inserted in a groove cut in the two adjoining surfaces of a joint. The groove is cut with a saw blade or dado head to a specific width and depth. (The groove for the spline is commonly run in with the dado head, ¼ inch being usual for ¾- to 1-inch stock, although a ⅛-inch spline, a single saw cut, is sometimes used, especially for miters.) A thin piece of stock is then cut to fit into this groove. The spline stock should be cut so that the grain runs at right angles to the grain of the joint.

A very simple way to produce splines is to cut up scrap pieces of ⅛-inch plywood. A supply of these can be kept on hand. The advantages of the plywood is its strength in each direction and its constant thickness. Quite probably your saw blade cuts a ⅛-inch kerf which is just right.

Clamp nails can be used as splines. These nails are flat with a flanged edge that fits into the spline cut. They can then be driven with a hammer to hold the two parts firmly together.

Other Wood Fasteners

There are several other fasteners on the market which are good for holding together wood joints. The best known are the corrugated fasteners and chevrons.

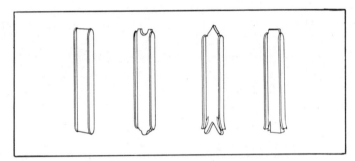

Typical clamp nails.

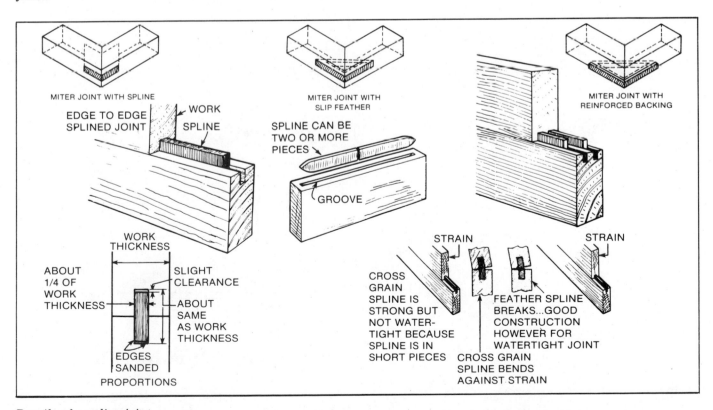

Details of a spline joint.

Corrugated Fasteners   The corrugated fastener, or wiggle nail, is used particularly in the miter joint. It is made of sheet metal of 18 to 22 gauge with alternate ridges and grooves; the ridges vary from $\frac{3}{16}$ to $\frac{5}{16}$ inch, center to center. One end is cut square; the other end is sharpened with beveled edges. There are two types of corrugated fasteners: one with the ridges running parallel and the other with ridges running at a slight angle to one another. The latter type has a tendency to compress the material since the ridges and grooves are closer at the top than at the bottom. These fasteners are made in several different lengths and widths. The width varies from $\frac{5}{8}$ to $1\frac{1}{8}$ inches, while the length varies from $\frac{1}{4}$ to $\frac{3}{4}$ inch. The fasteners also are made with different numbers of ridges, ranging from three to six ridges per fastener. Corrugated fasteners are used in a number of ways: to fasten parallel boards together, as in fastening tabletops; to make any type of joint; and as a substitute for nails where nails may split the timber. The fasteners have a greater holding power than nails in small timber.

When driving corrugated fasteners, use a medium-weight hammer and strike with evenly distributed light blows. It is important that the lumber being fastened together rests on a solid surface.

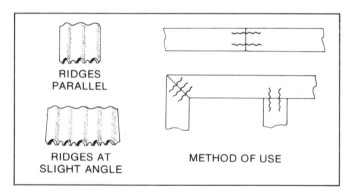

Corrugated fasteners and their uses.

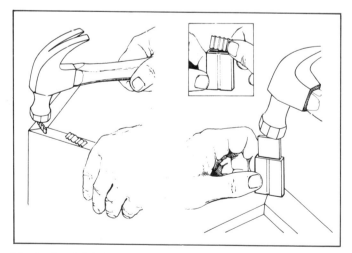

Methods of driving corrugated fasteners.

Chevrons and Skotch Fasteners   Like corrugated fasteners, chevrons and skotch fasteners are designed to draw wood together to make a tight joint. They are also driven in the same manner as corrugated fasteners.

Another interesting fastener is the barbed dowel pin. It has many functions such as aligning parts, serving as a pivot point, and permitting easy disassembly or separation of items.

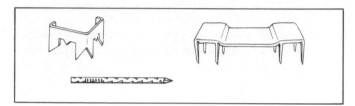

A typical chevron, skotch, and dowel pin fastener.

Tee-Nuts   These nuts are made in several different types, shapes, and sizes, but they basically perform the same job. While they are suited for many joining tasks, they are often used in conjunction with hanger screws to join legs to other wooden furniture pieces.

To install a tee-nut, drive it into the wood piece that is to hold the jointed portion. Make sure that the base with its twisted prongs is flat against the wood surface. If a hanger screw is used, the woodscrew end is installed in the end of a leg, then the bolt-threaded end is fitted into the tee-nut. For other installations a small bolt or machine screw can be used.

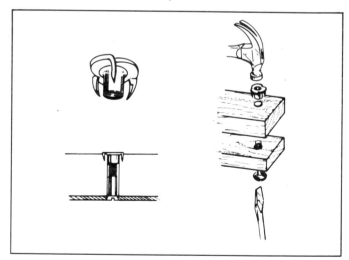

A tee-nut and how it is installed.

Braces and Mending Plates

Metal fasteners, generally called angle irons and mending plates, can be used to reinforce wood joints, especially when doing repair work. Available in many shapes and sizes, they can be set either into the wood or on its face. A mending plate recessed into the wood has more holding power than one placed on the surface. One half is first centered on one member of a joint and screwed in place. The second member of the joint is then held tightly against the first, and the other half of the fastener is screwed fast.

Corner Braces or Angle Irons Corner braces of the proper size can be used to support any two pieces of wood when they are attached at 90 degree angles. Corner braces can be used to give extra support for table legs, chair legs, or in almost any location where one piece of wood is joined to another at this angle.

You can set a corner brace more tightly into position by using a thin piece of cardboard. One flange of the corner brace is first affixed to one piece of wood, while the piece of cardboard is held tightly between the other flange and the adjoining piece of wood. Fasten the first flange to the wood as tightly as possible; then remove the cardboard and insert the screws in the opposite flange. The small piece of

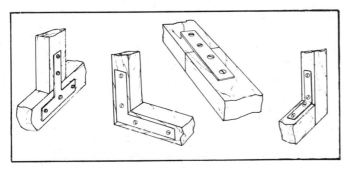

Common types of braces and mending plates.

cardboard helps to keep tension on the screws originally set. When it is removed, the space the cardboard occupied provides just enough leverage to tighten the additional arm more securely than could be done without the cardboard.

Larger corner braces can be used as shelf supports

for small shelves. If properly positioned and spaced, they will enable an ordinary shelf to carry a great deal of additional weight without sagging. When using angle irons as wall shelf supports, they should first be attached to the studs in the walls. That is, after the studs are located, screw the corner brace into the stud first; then drill a starting hole and insert screws of the proper length into the bottom part of the shelf.

Although corner braces are made to fit two pieces of wood joined at a 90-degree angle, they can be bent to fit wood set at other angles. Small corner braces can easily be bent with two pair of pliers. Larger corner braces may have to be bent with a vise and larger pliers. To bend the corner brace to the exact angle desired, it is suggested that you trace the angle you want on a plain piece of paper, and then bend the corner brace until it conforms to this sketched angle. If bent with care, a corner brace can be altered to almost any desired angle.

Wider corner braces are available that can be used where wider pieces of wood are joined. These are generally designed for use on inside corners. They can be attached with regular wood screws or with small lag screws where additional strength is required.

Corner braces that are neatly designed and copper- or chrome-plated are made for use where they might be exposed to view. Although small in size, they can be attached to chair legs, furniture legs,

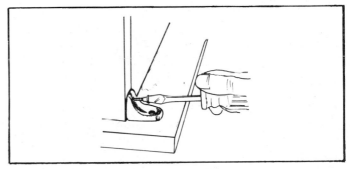

Copper- and chrome-plated corner braces are available for use where they may be seen.

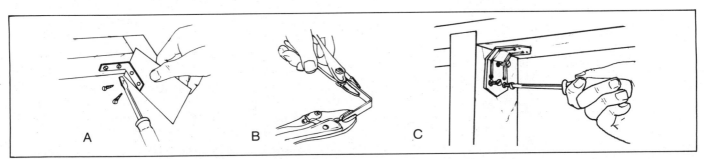

(A) Corner braces can be fitted tightly by inserting a piece of cardboard, fastening the opposite flange, then removing the cardboard. (B) Corner braces can be bent to fit any *angle. (C) Corner braces with extra width may be ideal for certain jobs.*

and so on to provide additional support without creating an unsightly appearance. In most cases, even the copper-plated braces are not exposed to view; but, if they are seen, they are still not at all unsightly.

Special tabletop fasteners are made for attaching tabletops and chair seats to the frame of a table or chair. These are special fasteners designed exclusively for this purpose. The fastener itself is attached to the underside of the chair seat or tabletop. It has a slight lip which then fits into a kerf or slot that is sawed around the apron of the chair or table. If special tabletop fasteners of this type are not available, the job can be done with a corner block.

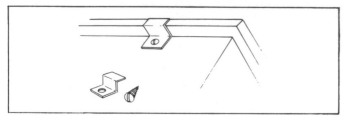

Special tabletop fasteners fit under the table for extra support.

T-Braces and Mending Plates   Mending plates run from small to large. They can be attached to small pieces of wood with light screws, or larger sizes of plates can be used and attached to the wood with small lag screws. In some cases, mending plates are even attached to wood with carriage bolts. Right-angle and T-shaped mending plates form strong joints when used at joints without glue or nails. When a joint is screwed, nailed, or glued and then supported by a T-shaped or right-angle mending plate, it becomes almost unbreakable.

Although T-braces and mending plates are designed for use on flat surfaces, they can be bent to conform to offsets in wood. If a T-plate brace is to be bent, the exact length and angle of each bend should be carefully measured on the brace and clearly marked. The T-plate can then be placed in a vise and

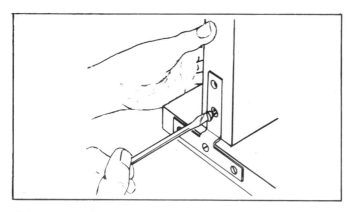

T-braces or plates can be bent to fit any holding job.

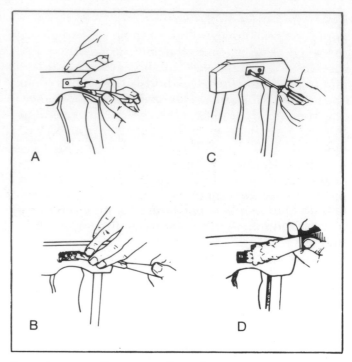

Steps in concealing a mending plate.

bent with a vise-grip pliers or any other strong holding wrench. T-plate braces of any size can be bent for offset holding patterns similar to this. Smaller T-braces should be attached with screws. Larger T-braces can be attached with small lag screws or carriage bolts.

In some cases, you may want to use mending plates for repairing broken pieces of wood but you may also want to conceal the plate. The first step in achieving this task is to position the mending plate on the broken area and mark exactly where it is to be attached. Use an ordinary pencil or a scratch awl for marking these dimensions on the wood. After the exact location for the mending plate has been carefully marked, select a sharp wood chisel and begin to mortise out the wood deep enough to embed the mending plate. Use the sharp chisel to make cuts along the marked lines. Use a plastic or wooden hammer to drive the chisel into the wood along the marked lines. Cut to the desired depth of the mortise. The depth will depend on the thickness of the wood being repaired.

The next step is to smooth out the bottom of the mortised area by holding the chisel flat with the beveled side up and rubbing it lightly back and forth over the mortised area. Use extreme care. Hold the cutting blade within the confines of the mortised area at all times while cutting. Also be sure to use a sharp chisel of the correct size.

When the area to be repaired has been thoroughly mortised out, you are ready to screw the mending plate into position. The plate should be embedded at least ⅛ inch below the surface of the wood. Lay the plate into the mortised position to make sure it fits smoothly along the bottom of the mortised area. If there are rough spots, remove the mending plate and sand or smooth these spots away. When you have smoothed out all the rough spots so there is no rocking of the plate on an uneven bottom, screw the plate firmly into position. Be sure to use screws of the correct length so they will not penetrate the wood on the opposite side.

The final step in concealing a mending plate is to fill the mortised area with water putty or wood plastic. Use a putty knife to pack the filler material into the mortised hole. Fill the mortised area up to the surface level of the surrounding wood. Pack tightly into the mortised spot and remove any surplus filler material with the putty knife. Use a fine grade of sandpaper to smooth the patched area. After it is carefully smoothed and evened to conform to the regular contour of the wood area, it can be refinished to match the original wood.

Structural Wood Connectors

Up to this point in the chapter we have concerned ourselves primarily with strengthened joints used in finishing woodworking. Let us take a look at some of the more common wood connectors that take the fuss out of framing. Each type should be installed according to manufacturer's instructions.

Anchor Clips   The use of anchor clips is an improved approach to anchoring wood to masonry because they eliminate the time-consuming, difficult task of locating and drilling holes for anchor bolts.

Post Base Clips   These structural connectors offer fast, adjustable, and correct installation of 4 by 4-inch posts to concrete. The slotted base can be installed even when anchor bolts are bent. The base can also be attached with concrete nails through the bottom holes. The termite- and rotproof base cover has weep holes to allow moisture to escape.

Beam Clips   These fasteners provide a good

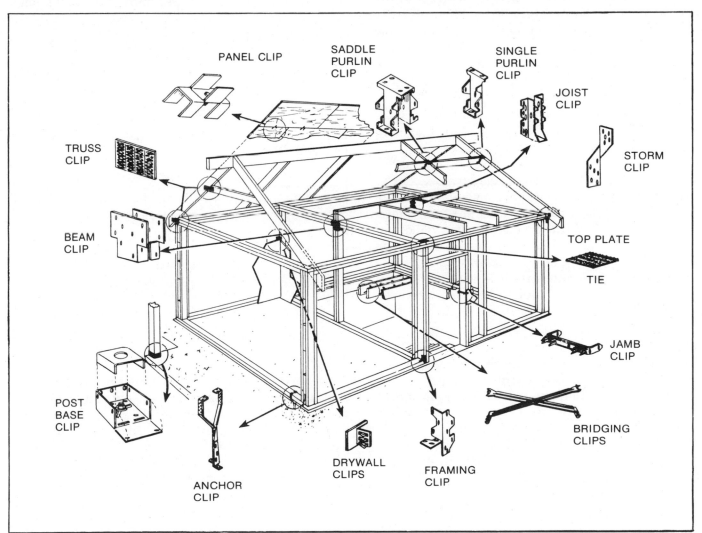

Structural wood connectors at work.

appearance and structural strength for anchoring the top of a post or column to a beam.

Bridging Clips   These clips save time and money by eliminating the need to cut, miter, and nail conventional wood bridging. They can be installed between 16 inches on center joists after a plumbing, wiring, or ductwork operation.

Framing Clips   These clips make short work of all two- and three-way ties. One clip handles roof, floor, and wall framing.

Storm Clips   These clips provide the proper method to eliminate wind uplift problems by anchoring trusses or rafters securely to plates and studs.

Joist Clips   Joist clips ensure that ceiling and floor joists are properly attached to headers. Joist clip's integral speed-nailing and positioning tabs allow rapid installation.

Purlin Clips   These structural wood connectors can speed up framing, layout, and installation. By using the purlin clips to install your roof, purlins can be placed between trusses, flush with the top of each truss. This eliminates the problem of birds nesting between trusses and roofing, improving cleanliness.

Panel Clips   These clips reduce costs on plywood roof sheathing. They eliminate costly edge blocking, and their special design provides for a tight, snug fit.

Dry Wall Clips   Structural dry wall fasteners can cut framing and lumber costs by as much as 75 percent. They eliminate two 2 by 4-inch studs in each corner as well as ceiling back-up material. For example, if the dry wall clips are fastened 16 inches on center—the maximum distance they should be spaced—only 250 of them would be required on an average 1,250 square foot area, saving fifty studs and 145 lineal feet of back-up materials.

The dry wall clip's exclusive integrated speed nail design cuts installation time to less than one minute per corner.

Jamb Clips   Jamb clips are a fast, simple way to install prehung or job-hung doors. Because they are self-nailing, they eliminate shimming or wedging of the door. Also, there are no nail holes or hammer marks to sand on the face of the jamb. Eight clips are needed for hollow core doors and ten for solid core.

Top Plate Ties   The top plate tie is a better and faster way of connecting walls and partitions. They eliminate the measuring, cutting, and notching re-

quired by the conventional fly-by and hold-back method. An average top plate tie holds over a half ton, which far exceeds the strength of the 16d nails applied with a hammer.

Truss Clips   Truss clips are usually made of 18-gauge, quarter hard, zinc-coated steel. Their holding power is about 50 pounds per tooth.

Fence Brackets

Designed to simplify putting together fences, fence brackets provide a more efficient connection than can usually be obtained when using conventional framing methods. They are useful in building a wide variety of fence patterns: straight rail, louvered rail, conventional picket, louvered screens. Fence rails are easily and quickly connected to posts by nailing or screwing without the need for special cutting, notching, or other fitting. The slide-in feature permits the simple removal of entire fence

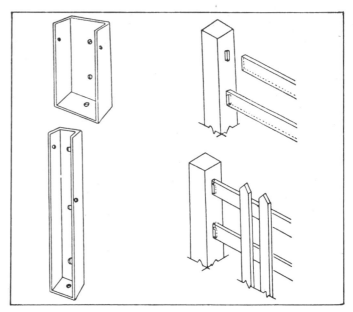

Fence brackets designed to simplify putting together fences.

sections or individual rails and louvres so that maintenance work, lawn cutting, painting, or snow removal is more easily accomplished. They are also extremely useful in making special screening louvres (both horizontal and vertical patterns) for patio and porch privacy, carport enclosures, and contemporary interior partitions.

Other Ways to Put It All Together

While the major put-it-together materials have already been discussed, there are several others that are important for certain types of fastening and holding jobs. These materials include tape, rope, and hinges, plus a few other holding items.

Tape

There are many types of tape that help make putting it all together a great deal easier. Of course, tape performs other tasks, such as binding, masking, mending, and decorating. Here are some of the more common types as well as specialized ones, together with possible uses.

Masking Tape A quality masking tape unwinds easily without splitting. Considered a pressure-sensitive tape, it has excellent ability to stick immediately and securely to nearly all surfaces yet pulls away without damaging the surface. Because of this, it can be used to mask out parts of a surface to be left unpainted on painting jobs, to hold pieces of wood together while checking an assembly of cut parts before gluing, to prevent the surface of a plywood board from splintering when sawed, to tape a nut to your wrench to start it on hard-to-reach bolts, and to avoid wrench marks on chrome fixtures.

Double-Faced Tape This pressure-sensitive tape has high holding power on both sides. It can be used to put things on the wall, such as lightweight pictures and plaques, or to fasten photos in an album. It is also good for securing carpeting, runners, rugs, and underlay. Its double face adheres securely to both fabric and flooring, forming a bond that prevents creeping, bulging, and overlapping. Actually, this tape has many interesting uses.

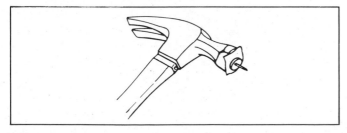

Convert your nail hammer to a tack hammer by coating the head with double-faced tape.

Cloth or Fabric Tape This is another pressure-sensitive tape in several different colors. It can be employed to mend fabrics, to repair book binding, to patch auto seat covers, and for many other household put-it-together tasks.

Filament or Strapping Tape Reinforced with glass fibers, this tape is as strong as wire, but is very flexible. It is used in place of clamps for holding awkwardly shaped parts when the pieces have been glued. It can also be used to mend broken handles of garden tools or to bundle almost anything: lumber, pipes, pieces of molding, ropes, hoses, and so on.

Utility or Duct Tape This heavy-duty tape is designed to keep your heating and cooling system at top efficiency. Wrap it where the ducts are joined to seal in heat. It is also useful to repair canvas, seal heavy packages, or to seal around a window-installed air conditioning unit. Some utility tapes are available that consist of asphalt adhesive with an aluminum facing, which makes it good for general repair in roofing, guttering, leaking pipes, and hoses.

Sponge Tape This pressure-sensitive tape is available for padding and absorbing shock. It is ideal for bottoms of lamps, stereos, ashtrays, and bookends, and is more durable than felt. Also, sponge tape can be used to weatherstrip doors and windows; it cuts down on shock, deadens vibration, and insulates.

Plastic Repair Tape This vinyl, waterproof tape can be used to repair inflatable toys, beach balls, wading pools, air mattresses, hoses, umbrel-

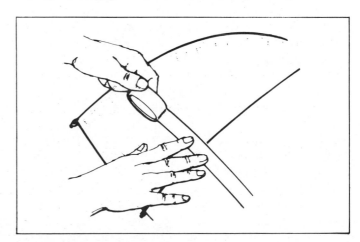

Patching an umbrella with clear plastic repair tape.

las, and shower curtains. Because it is available in various colors as well as clear, the tape can be used to make a tough, colorful border for plain vinyl material.

Plastic Electrical Tape   Flexible and tough, it has replaced the old standards—rubber and friction tape—for insulating electrical wiring. It has certain advantages over rubber and friction tape. For example, it will withstand higher voltages for a given thickness. Single thin layers of certain commercially available plastic tape will stand several thousand volts without breaking down. However, to provide an extra margin of safety, several layers are usually wound over the splice. Because the tape is very thin, the extra layers add only a very small amount of bulk; at the same time, the added protection, normally furnished by friction tape, is provided by the additional layers of plastic electrical tape. In the choice of plastic tape, expense must be balanced against the other factors involved.

Apply the splicing tape smoothly and under tension so that there will be no air spaces between the layers. In putting on the first layer, start near the middle of the joint instead of the end. The diameter of the completed insulated joint should be somewhat greater than the overall diameter of the original cable, including the insulation.

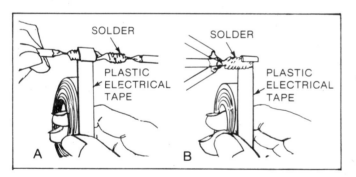

(A) Taping a soldered joint. (B) Taping a bunch splice.

In addition, plastic electrical tape has many other uses, including making a quick repair on a plastic garden hose, insulating handles on pliers and screwdrivers, and repairing frayed ignition wires in a car.

Wood Tape   A wood veneer tape is available for gluing to the end grain of plywood panels in order to conceal the ply construction. It is available in various wood grains and can be applied with any of a number of adhesives. One kind has adhesive backing and is pressed on with a warm iron.

Velcro Tapes   The Velcro hook-and-loop tape ("vel" for velvet and "cro" for crochet) is used in hundreds of household and industrial jobs. You may already have clothing that uses Velcro fasteners. To use them, you simply press to close, pull apart to open. Velcro has many other uses around

The hooked portion of the tape has a great number of precisely shaped snags (left). When the hooked portion is pressed against the looped piece (right), fastening occurs, even if they are not parallel. No fastening will occur if you press two loop or two hook pieces together.

the home, shop, car, or boat and can probably make a lot of your put-it-together tasks easier.

Velcro is available in two materials—nylon and polyester—and in three types of backings—noncoated, coated, and pressure-sensitive. The nylon fasteners are recommended for use where they will not be exposed to a great deal of moisture and sunlight. Where these two conditions exist, use polyester Velcro. The noncoated type is intended for sewing application, while coated-backed is suggested where gluing is the desired method of fastening. Any adhesive recommended for cloth can be used with Velcro. The pressure-sensitive backing is good for quick applications. For best adhesion, the area must be clean. Even then, this pressure backing should not be considered where a permanent bond is desired. It will, sooner or later, break loose from the surfaces. The coated and noncoated types can also be attached with nails, screws, or staples.

These tapes are usually available in two widths: ¾ and 1½ inches. The amount of tape required for any given job is directly proportional to the strength needed. Four square inches of Velcro has twice the holding strength of two square inches. The exact amount you require is dependent upon where the tape is to be employed and the forces that will be acting upon it. Experimentation and a little common sense are good guides when selecting the amount of tape needed.

Other Types of Tape   There are many special tapes available for unique putting-it-together jobs. For instance, book tape, available in a full range of sizes and colors, is ideal for repairing book bindings and covers. There are package tapes that are strong and self-sealing for mailing and wrapping. For preparing food items for a freezer, there is a special freezer tape that holds tight even to –40 degrees F. Reflective tape glows when light hits it at night and can be used on the rear bumper of a car, on the wheels of children's bicycles, on markers along the driveway, and even across the stairs in dark places. Finally, there are clear plastic or cellophane tapes that perform so many tasks it is virtually impossible to name all of them.

Rope

Rope is a common put-it-together material. Today the most popular ropes are synthetics—nylon, Dacron, polyethylene, and multi- and mono-filament polypropylene. Manila hemp, sisal, and braided cotton ropes are still on the market, but on a smaller scale than previously.

All synthetics are virtually impervious to weather, rot, and mildew, although some polyethylene and polypropylene will break down under prolonged exposure to sun. To various degrees, they are resistant to abrasion, acids, and caustics. With the exception of Dacron, all synthetics are lighter than Manila hemp. Without exception, all of the synthetics are stronger than Manila.

Constructions vary according to the way in which the strands are twisted into rope. Most common types are three-strand, four-strand, braided, and plaited. Plaited rope is fairly new. Unlike three-strand rope, eight-strand plaited rope will not turn while in use. When not in use, it is completely relaxed and does not wear out as fast.

While braided rope is limited to the eye splice only, plaited rope will actually work with a quick temporary pseudo-splice if necessary. Unlike the braided product, the direction of twist of the yarns in plaited rope does not cause it to snag; it will not flex open to permit entry of dirt or abrasive material; it can easily be inspected internally for signs of wear; it offers a better gripping surface.

Manila Rope   This rope is made of carefully selected natural fibers and is an excellent general purpose rope for maritime, industrial, and agricultural uses. It is often waterproofed.

Nylon Rope   Nylon rope, which is 2½ to 3 times stronger and will last 4 to 5 times longer than natural fiber ropes, is available in twisted, solid braided, and braid-on-braid constructions. Twisted has been a favorite for many years because it is stronger than solid braid and can be spliced. A solid braid handle is much better than twisted because it does not kink or knot under strain. Its main disadvantage is that it cannot be spliced.

Braid-on-braid nylon rope, which has a braided core inside a braided jacket, combines the best features of both twisted and solid braid constructions. It handles well, is strong, and can be spliced.

Polypropylene Rope   It is strong, lightweight, easily handled, and floats on water. It has a better "hand" or "feel" than polyethylene and is slightly stronger. It is available in twisted, solid braid, and diamond braid constructions. The latter braid has the advantage of almost instant splicing. Eye splices, end-to-end splices, and crown splices can be made in short time.

Polyethylene Rope   It is in the same family of fibers as polypropylene but is a somewhat slicker rope, and as a result, has lost some of its popularity to polypropylene. It is used as a barrier rope and for general purposes in home, farm, and industry. It is available in the same constructions as polypropylene rope.

Polyester or Dacron Rope   This type of rope is ideal for a running rigging, bolt rope, and other applications where a minimum of stretch, high strength, and durability are needed. Its greater strength (size for size, 2 to 2½ times as strong as Manila rope), low moisture absorption, and resistance to wear, all combine to give long service life.

Combination ropes are available which pair polyester with either polypropylene or polyethylene. These ropes combine desirable qualities of each fiber. They are normally used as all-purpose, low-cost rope.

Wire Rope   Available in various sizes, this rope is used for boat and auto trailers, winch lines, antennae, tree guides, overhead door cables, and other similar applications.

Twine   Household twines, which are frequently used to put things together, generally include wrapping, parcel post, express and kite twines. Most are made of cotton, but jute and sisal are still sold. Polypropylene twine is the newest form of twine. It is popular for tying packages and bundles because it is strong, rotproof, and available in colors. Heavier twines, known as seine twine or cable cord, are made of cotton or nylon.

Hinges

While many people may not consider hinges as fasteners, they do play an important part in putting together many items. A hinge, of course, is a mechanical device consisting primarily of two plates and a pin. One plate is attached to the door and the other plate is attached to a fixed surface, either a cabinet side or a door jamb.

The hinge required for any job depends on the design of the door and frame, the size and weight of the door, and the amount of traffic expected to use the door. For example, a standard house door between 60 inches and 90 inches requires three hinges. The range in hinge sizes extends from about 3 inches in height for a narrow, hollow core door to 5 inches for a solid core door, 2 inches thick by 36 inches wide. The specific type hinge used depends on the construction of the door and frame and the general appearance desired. Let us take a look at some of the more popular hinge types.

Butt Hinges   A butt hinge fits between the butt of the door and the frame with only the hinge pin exposed on the inside of the door.

A full mortise butt hinge is mortised into both the door and the frame. Both leaves of the hinge are swaged so that the surfaces come together when the hinge is closed. Some butt hinges contain ball bearings which are permanently lubricated. They are good for heavy doors opening to the exterior.

Full-Surface Hinges   A full-surface hinge is mounted with both leaves on the surface of the door and frame so that the entire hinge is exposed when the door is closed. Neither leaf needs to be mortised. Both leaves are flat so that the hinge lies flush against the door and frame.

Half-Surface Hinges   These are mounted with one leaf mortised and the other surface applied. A half-mortise hinge is similar except the surface applied leaf is narrower than on a half-surface hinge.

Raised Butt Hinges   These hinges are used when heavy carpeting interferes with the opening of the door. As the door is swung open, it rises slightly to clear the carpeting.

Pivot Hinges   Available in various designs, the gravity pivot hinge mounts at the top and bottom of the door with only a small wafer of metal exposed to view. They are commonly used on furniture doors or in situations where doors are intended to be inconspicuous.

Spring Hinges   Spring-loaded hinges bring the door to a closed position automatically. The double-acting type are commonly used on cafe doors where doors swing in either direction.

Friction Hinges   These hinges are designed to swing a door and hold it at any desired position by means of friction control incorporated in the knuckle of the hinge.

Strap Hinges   Specifically for surface applications, the extended leaves of these hinges provide greater support, especially for wide doors. The popularity of the design has resulted in the development of decorative strap hinges.

Continuous Hinges   Commonly called piano hinges, these range up to 84 inches in length and are used for continuous application along the entire length of the door. They provide maximum strength and protection against warping, making them popular for chest lids, cabinets, and so on.

Invisible Hinges   There are several types of invisible hinges on the market, and most can be used in many places within the home—for room doors, cabinet doors, or secret panels. No portion of the hinge is visible when the door is closed, thereby leaving the entire outer surface of the door and frame free from any projections and for any desired

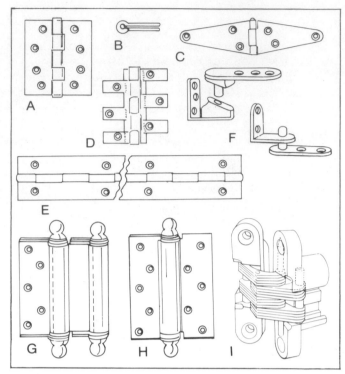

Various types of common hinges: (A) butt hinge, (B) full-mortise with swage-top view, (C) strap hinge, (D) non-mortise butt hinge, (E) continuous hinge, (F) pivot hinge, (G) double-acting spring hinge, (H) single-acting spring hinge, and (I) invisible hinge.

treatment or decoration. These hinges are made for all types of doors no matter what their size or weight. The invisible hinge allows the door to open fully 180 degrees, and since it is not visible, it can be used without concern about its matching other hardware.

Gate Hinges   While several of the hinges already described can be used on gates, there are some designed especially for gates. When installing any gate hinges, be sure to follow the manufacturer's recommendations.

Cabinet Hinges   There are three basic cabinet door designs that determine the type of hinge required in any situation: flush mounted, lipped, or flush overlay.

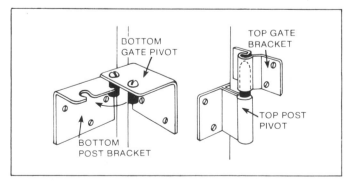

Two types of gate hinges.

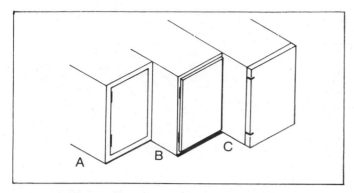

Methods of setting cabinet and furniture doors: (A) flush, (B) lipped, and (C) flush overlay.

Flush mounted doors can use full-mortise butt or full-surface hinges, ornamental strap hinges, and concealed hinges, much the same as any standard house door, since the cabinet door is recessed into the frame.

Lipped doors are partially recessed into the opening with a lip extending around the outside of the frame. In this case, the hinge must be offset to accommodate the lip. Semiconcealed cabinet hinges are designed so that the leaf attached to the cabinet frame is exposed and the leaf attached to the door is concealed. This means that the hinge must be offset into the closing side by the thickness of the door that is recessed into the cabinet.

A surface hinge for a lipped door must be offset to the outside of the door. Here the offset must match the thickness and shape of the portion of the door that extends outside the cabinet opening.

Most lipped cabinet doors have a ⅜-inch offset, but you should measure to be sure you are buying the right hinge. Flush overlay doors (doors that completely overlay the cabinet frame) are usually mounted with pivot hinges mortised into the top and bottom of the door, butt hinges, or a semiconcealed hinge with no offset on the door leaf.

A self-closing feature can be built into various kinds of hinges. Most work on a spring principle. This is a quality feature that appeals to housewives who have problems with children leaving cupboard doors open.

Basic material, construction, and finish are elements of quality in any hinge. A good cabinet hinge should have a five-knuckle joint for better load distribution and smoother, quieter action. Pivot hinges should have firmly riveted joint pins and nylon washers between the hinge wings.

The material from which the hinge is made should be sufficiently heavy to prevent sagging and should be resistant to normal kitchen moisture.

Desk-Top Hinge Supports   When putting together furniture, you may want to provide for a writ-

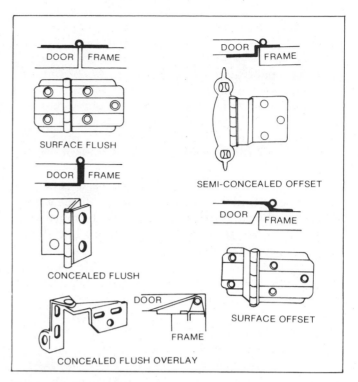

Typical cabinet hinges.

ing surface or desk top that is not visible when the cabinets are closed. The basic unit is similar to any other type of cabinet construction. The door, instead of being hinged at the side, is hinged at the bottom.

Regular butt hinges or a continuous hinge strip is fastened to the cabinet base and the lower section of the door. When the door opens forward and downward, it becomes a desk top. To keep the desk top door level with the floor, support braces are added. These are made of steel or solid brass and are available in many different styles. A support should be placed on each side of the desk top to help keep it level.

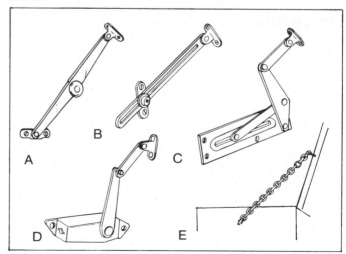

Various types of lid-supporting devices: (A) latch, (B) friction catch, (C) self-balanced, (D) counterbalanced, and (E) chain.

Knobs, Pulls, and Catches

Knobs, pulls, and cabinet catches play a very important part in putting furniture and cabinets together.

Knobs and Pulls Knobs and pulls are used on cabinet doors and drawers as handles. The basic consideration to the customer is usually style, but quality and design can be important. Backplates are available for many models and help protect the surface of the cabinet as well as add a decorative touch. Various other types of drawer pull ideas are available.

If the hardware is replacing previous equipment on the cabinet, consideration must be given to either using existing screw holes or being sure that the new hardware will cover the old holes. A two-screw cabinet handle, for instance, cannot be replaced with a single-screw knob, unless a backplate is included and is large enough to cover the second hole.

Catches Normal wear and settling will cause almost any cabinet door to sag after a few months or years. When this happens, the doors will hang open unless catches or self-closing hinges are installed. Catches come in six basic types—friction, roller spring, magnetic, elbow, bullet, and touch (push) catches.

Friction catches hold by pressure of the catch on the strike. The catch is mounted on a door frame, jamb, or underside of a shelf. The strike is mounted on the door so that upon closing it is inserted into the catch. The two most common friction catches are alligator and lever spring-action, which features two floating jaws and is self-aligning to compensate for swelling and shrinking of doors.

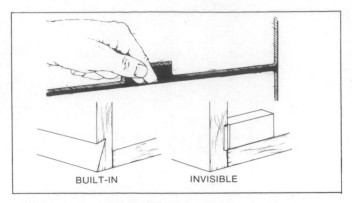

Other drawer pull arrangements.

Roller spring catches are available in single and double roller types. They feature quiet operation, easy installation, long life, and easy adaptability to many door and frame designs.

Magnetic catches range in pull from 8 to 10 pounds. Since holding power is greatly reduced if only part of the magnet makes contact with the strike, they must be installed carefully to properly align the catch and strike. Quality magnetic catches feature a floating or self-adjusting action to insure proper alignment and contact.

Elbow catches are the opposite of other catches in that they are mounted on the door with the strike installed on the frame or on a shelf. These catches can only be released from the inside of the cabinet and thus should only be used on one side of a pair of doors.

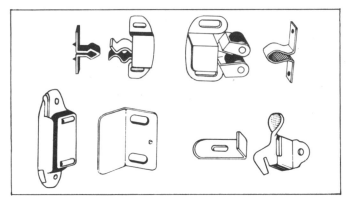

Several common door catches.

Bullet catches are used primarily on furniture and smaller cabinet doors where it is desirable to hide the catch as much as possible.

Touch (push) catches are mounted inside the cabinet and have no knobs or pulls. By simply pushing on the door, the catch releases and the door springs open.

Hasps and Staples These are door fastenings that are really a combination hinge and catch. They can be secured by a padlock or peg.

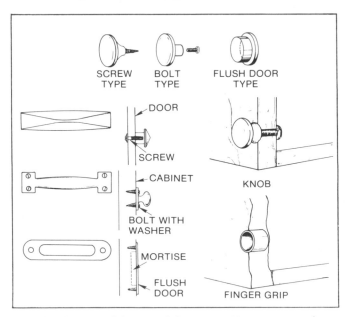

Standard types of door and drawer pulls.

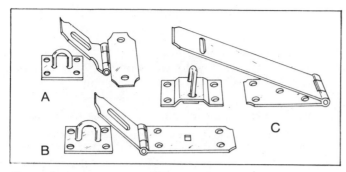

Several common hasps and staples: (A) hinge hasp for chest, (B) hinge hasp, and (C) safety hasp.

Gate Latches A gate latch can be anything from a simple hook-and-eye to the special types shown.

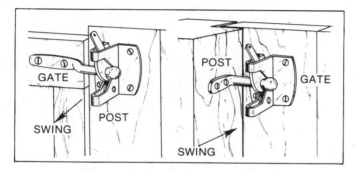

Gate latches for gates that swing only one way.

Drawer Slides

To put drawers together with their furniture unit, you need some type of drawer slide. You can make your own slides or purchase them ready-made. For example, an ordinary shelf can support the drawer which will slide on the bottom edge of its sides. While this method works, you are likely to encounter problems if the wood swells or the drawer warps.

Dadoes cut in the outside face of both sides can be used as guides, permitting the drawer to ride on cleats. A matching cleat is glued to the side of the drawer. In either case, these arrangements keep the weight off the bottom. If the cleats are made of hardwood and are finely finished and waxed, you should encounter little difficulty for a period of time.

To avoid cutting dadoes in the drawer or cabinet sides, you can add two small cleats to the shelf on which the drawer rests and screw a wider board into the base of the drawer. This serves as a guide between the nailed cleats. It will help keep the drawer straight, but may eventually cause trouble in opening and closing.

Metal slides are the surest way to trouble-free drawer operation. Two basic types of metal drawer

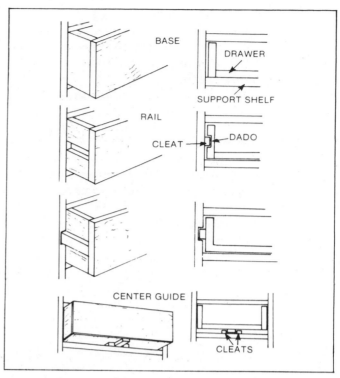

Several shop-type drawer slides.

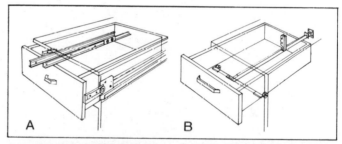

Two basic types of metal drawer slides: (A) side-mounting and (B) monorail or underdrawer mounting.

slides are side-mounting and monorail, with various models available in each type.

The side-mounting version consists of tracks, one attached to each side of the drawer and one on both the left and right side of the cabinet. Each track has integral rollers on which the drawer rides. To install a side mount, the drawers must usually have ½-inch clearance on both the left and right sides. Slides may not be installed on drawers with a height of less than 3 inches overall. The slides themselves are put in place as follows:

1. Fasten channels marked C-R and C-L (cabinet right and cabinet left) to the cabinet. The channel should be flush with the front and bottom of the cabinet opening.
2. Fasten channels marked D-R and D-L (drawer right and drawer left) to the drawer sides. The roller plate should be flush with the drawer bottom.

3. Insert the drawer, now ready for use. Drawers are easily removed by pressing the fingertip release when the drawer is in the open position.

Self-closing slides are also available in side-mounted design. These close when the drawer comes within 4 to 5 inches of the back, regardless of the load or its position in the drawer.

The monorail or underdrawer mount features a single track under the center of the drawer with drawer rollers on the left and right side. For best results, make the drawer ¼ inch narrower than the cabinet opening. The height of the drawer sides should be ⅜ inch less than the cabinet opening. The installation is made as follows:

1. Mark the vertical center line on the back of the drawer. Align this with the vertical scribe on the rear roller flange. Position the lower edge of the roller flange even with the bottom of the drawer back and fasten it with screws.

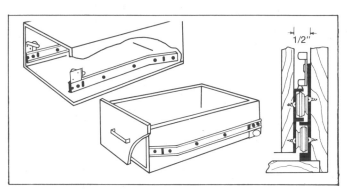

Installing a side-type drawer slide.

2. Slide the bracket onto the channel. Locate and fasten the channel in the opening, using the center line. Locate the side rollers with the flange against the sides of the cabinet opening.

3. Secure the back bracket on the cabinet center line loosely, using a screw in the slot. Insert the drawer and adjust the back bracket to align the drawer front when closed. Then tighten the screws.

Quality slides permit little side movement, prevent accidental drawer pull-out, have high quality rollers, and are precision made to close tolerances.

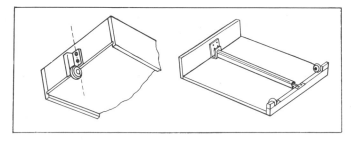

Installing a monorail-type drawer slide.

Shelf Supports

There are several shelf supports available to put shelves together in a bookcase or on the wall. When putting together a bookcase, you have your choice of constructing shelf supports or buying them.

One of the simplest methods is to use a cleat (½ by ½ inch, 1 by 1 inch, or quarter round) nailed or screwed to the side, with the shelf fastened to it. The cleats may be screened from direct view by installing vertical molding strips along the side edges of the shelves.

A stronger and neater way is the dado method of inserting shelves. A dado or groove is cut into each side, and the shelf is slipped into place. It can be held with nails, screws, or adhesive. When it is not desirable for the dado cut to show on the outside edge of the sides or uprights, you can use a blind or stopped dado.

Simple adjustable shelves can be made with short lengths of dowel to fit into holes drilled parallel to the edges of the sides or uprights. Under most circumstances, a ⅜-inch dowel should be used for a moderate-size shelf, but the exact size depends upon the length of the shelf and the weight it will support. In place of the dowels, you can use special metal shelf supports which fit into predrilled holes.

Possibly the simplest method of installing shelves is to use ready-made pilaster standards or strips. The standards should be installed on the surface or made flush with the wall or cabinet by cutting a dado.

When surface-mounting the standards, first mark the location of the strips according to your particu-

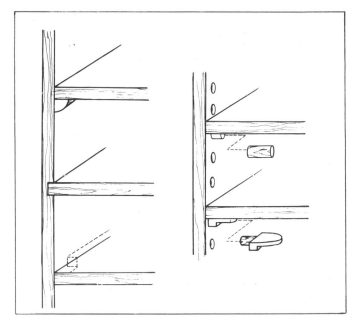

Shop methods of installing bookcase shelves.

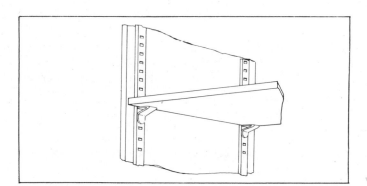

Surface-mounted pilaster standards.

lar needs. Nail the standards into the desired position at equal heights with the numbers on the standard pointing up. Check the horizontal and vertical alignments with a level.

When setting the standards flush, mark the location and widths of the strips, and rout the dadoes to the desired depth (about ⅜ inch). Place the standard in the dado and fasten it with the proper size of screw nails.

To install the pilaster clips, position the top hook into the desired slot and swing the clip downward until the tab engages the slot. It may be necessary to squeeze the clips slightly with a pair of pliers to make them fit. The shelves can be adjusted in height by moving the clips whenever desired.

Pilaster standards, when used in conjunction with shelf brackets, can be put on almost any wall surface. Because of their triangular design, these shelf brackets can support both wood and glass shelves nicely. The brackets are designed to fit pilaster standards and to handle 6 and 10-inch shelves.

To fasten the standards to the wall, use a ¾-inch long Number 4 screw with an appropriate wall anchor or fasten them to the wall studs. When the standards are installed, position the top hook of the bracket into the desired slot. Swing the bracket downward until the tab engages the slot. Once the brackets are installed, place the shelves upon them.

For a more decorative look, use standards and brackets which fit into slots in the standards. The

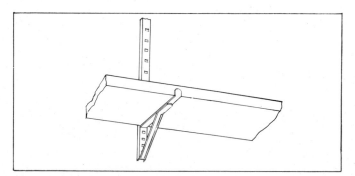

Shelf bracket in place on wall-mounted standards.

standards and brackets come in various sizes to meet practically every requirement. Furthermore, they are made in a polished aluminum finish, brass, wood, and in various decorator colors.

When positioning these decorative wall standards and brackets, remember that, as a rule, the heavier the load you want to put on your finished shelf system, the closer you must space the shelf standards.

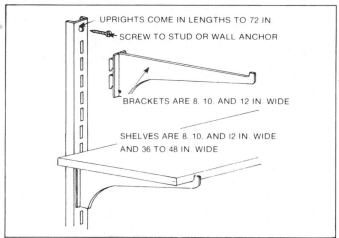

How decorative standards and brackets are installed.

If you are only going to store a few lightweight items on the shelves or make a decorative display of small items and a few books, you can space the standards from 32 to 36 inches apart. If you intend to keep heavy items on the shelves, such as large numbers of books, cans of paint, canned goods, and so on, standards should be no more than 24 inches apart.

Eight-inch-wide shelves are satisfactory for most books, although 10-inch shelves give a bit more elegant appearance to bookshelves. For other items, you should use the narrowest shelf that will conveniently meet the storage problem. This way, items on the shelves do not get buried behind other items, and the weight is distributed more evenly over the wall area. In any kind of shelf storage system, the idea is not to cram as many items as possible in a given shelf area, but to have enough shelves available to spread out the stored items so they can be seen and are easily accessible. The shelves can be an inch or two wider than the shelf brackets, if you wish. Just drill a ¼-inch hole in the underside of the shelf to receive the tip of the shelf bracket. The ends of the shelves should extend from 2 to 6 inches beyond the brackets, depending on the look you want to achieve.

Studs in the wall or wall anchors can be employed to hold the standards in place. When installing the brackets, it is important, after placing the shelf bracket in the standard, to insure that it is seated properly. Tap near the base of the bracket with a hammer or a screwdriver handle until it snaps in place.

Ready-made shelf brackets and angle irons can also be used to support cantilevered shelves. But, neither of these are adjustable or decorative.

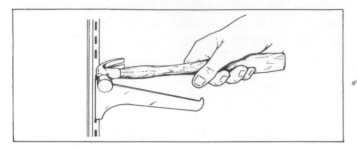

Tapping in a bracket.

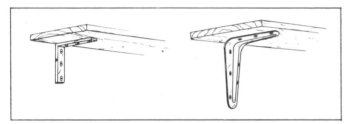

Ready-made steel brackets (left) and angle irons (right) can be used to hold shelves.

Electrical Connectors

Where there will be no strain on the wire, a quick and satisfactory method for joining wires is simply to use convenient solderless connectors.

Solderless Connectors These connectors are attached to their conductors by means of several different devices. Four of the most common types of solderless connectors, classified according to their methods of mounting, are the split-sleeve, split-tapered-sleeve, crimp-on splicers, and the wire nut.

To connect a split-sleeve splicer to its conductor, first insert the stripped wire tip between the split-sleeve jaws. Using a tool designed for that purpose, force the slide ring toward the end of the sleeve. Close the sleeve jaws tightly on the conductor, and the slide ring holds them securely.

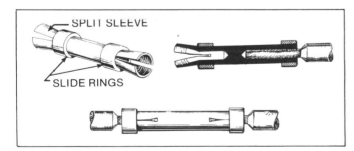

A split-sleeve splicer.

To mount a split-tapered-sleeve splicer, strip the conductor and insert it in the split-tapered sleeve. Turn or screw the threaded sleeve into the tapered bore of the body. As the sleeve is turned in, the split segments are squeezed tightly around the conductor by the narrowing bore. The finished splice must be covered with insulation.

The crimp-on splicer is the simplest of the ones discussed. Preinsulated and uninsulated types are manufactured. These splicers are mounted with a special plierslike hand-crimping tool designed for that purpose. The stripped conductor tips are inserted in the splicer, which is then squeezed tightly closed. The insulating sleeve grips the outer insulated conductor, and the metallic internal splicer grips the bare conductor strands.

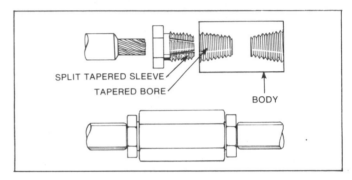

A split-tapered-sleeve splicer.

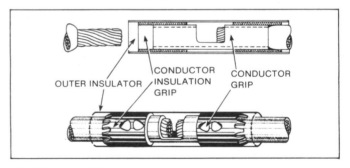

A crimp-on splicer.

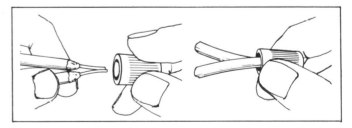

Method of installing a wire nut.

If there will be no strain on the wire, wire nuts may be used. Be sure no bare wire is left exposed.

Solderless Terminal Lugs Solderless terminal lugs are used more widely than solder terminal lugs. They afford adequate electric contact, plus great mechanical strength. In addition, they are easier to attach correctly because they are free of the most common problems of solder terminal lugs, such as cold solder joints and burned insulation. These lugs come in many sizes and shapes, each intended for service with wires of different size. Only the most common type—the crimp-on lug—is discussed here.

The crimp-on lug is simply squeezed or crimped tightly onto a conductor by means of the same tool used with the crimp-on splicer. When mounted, both the conductor and its insulation are gripped by the lug. For smaller solderless types of terminal lugs, a crimping tool is ideal.

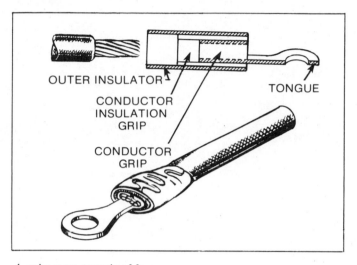

A crimp-on terminal lug.

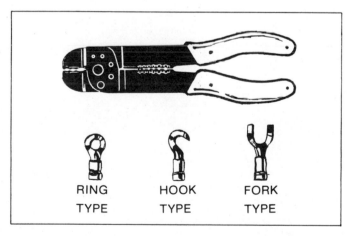

Crimping tool and other solderless types of terminals.

Fastener Storage

Fasteners should be carefully stored so that they will always be handy when you need them. Storage cabinets are readily available and some even come complete with a supply of the fasteners already in the cabinet drawers. Of course, there are several do-it-yourself schemes that you may devise. Baby food jars, coffee cans, and similar vessels also make good storage containers. Regardless of the container used, make sure that its contents are clearly marked.

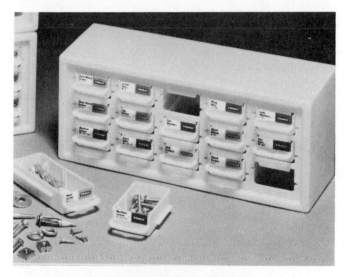

A typical ready-made fastener storage chest.

The ferris wheel is a good do-it-yourself project.

Index